D0990770

PLASTICS FOR ELECTRONICS

PLASTICS FOR ELECTRONICS

Edited by

MARTIN T. GOOSEY

Dynachem Corporation, Tustin, California, USA

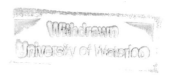

ELSEVIER APPLIED SCIENCE PUBLISHERS
LONDON and NEW YORK

ELSEVIER APPLIED SCIENCE PUBLISHERS LTD
Crown House, Linton Road, Barking, Essex IG11 8JU, England

Sole Distributor in the USA and Canada
ELSEVIER SCIENCE PUBLISHING CO., INC.
52 Vanderbilt Avenue, New York, NY 10017, USA

British Library Cataloguing in Publication Data

Plastics for electronics.
1. Plastics in electronics
I. Goosey, Martin T.
621.3815'1   TK7871.15.P5

ISBN 0-85334-338-1

WITH 51 TABLES AND 101 ILLUSTRATIONS

© ELSEVIER APPLIED SCIENCE PUBLISHERS LTD 1985

Printed in Northern Ireland by The Universities Press (Belfast) Ltd.

PREFACE

Much of the progress towards ever greater miniaturisation made by the electronics industry, from the early days of valves to the development of the transistor and later the integrated circuit, has only been made possible because of the availability of various polymeric materials. Indeed, many new plastics have been developed specifically for electrical and electronic device applications and as a consequence the plastics and electronics industries have continued to grow side-by-side.

Electronic components are one of the few groups of products in which the real cost performance function has declined significantly over the years, and part of the reason can be directly attributed to the availability and performance of new polymeric materials. The evolution of the personal computer is a specific example, where improvements in polymer-based photoresists and plastic encapsulation techniques have allowed the mass production of high-density memories and microprocessors at a cost which yields machines more powerful than mainframe computers of 30 years ago for little more than the price of a toy.

Today, plastic materials are widely used throughout all areas of electrical and electronic device production in diverse applications ranging from alpha particle barriers on memory devices to insulator mouldings for the largest bushings and transformers. Plastics, or more correctly polymers, find use as packaging materials for individual microcircuits, protective coatings, wire and cable insulators, printed circuit board components, die attach adhesives, equipment casings and a host of other applications.

This book is intended to serve as an introduction to the chemistry, properties and uses of plastics in electrical and electronic component processing and manufacture. No in-depth knowledge of polymer chemistry is assumed and it is aimed at the engineer who wishes to gain a greater understanding of the materials currently in use and to have a guide to the vast number of plastics that may be encountered. The book begins with an introduction to plastics, their structure and important properties, after which several chapters detail the major materials and uses. The very important topic of plastic encapsulated device reliability is also covered and the final chapter discusses a further number of materials and applications for plastics in the electrical and electronics industries.

Martin T. Goosey

CONTENTS

LIST OF CONTRIBUTORS

S. CLEMENTS

> Plessey Research (Caswell) Ltd, Caswell, Towcester, Northants NN12 8EQ, UK. (Present address: FTL, London Road, Harlow, Essex CM17 9NA, UK.)

J. COMYN

> School of Chemistry, Leicester Polytechnic, PO Box 143, Leicester LE1 9BH, UK

M. C. W. COTTRILL

> Plessey Office Systems PLC, Beeston, Nottingham NG9 1LA, UK

J. H. DAVIS

> Dow Corning Ltd, Barry, South Glamorgan CF6 7YL, UK

MARTIN T. GOOSEY

> Dynachem Corporation, 2631 Michelle Drive, Tustin, California 92680, USA

R. HURDITCH

> PPG Industries, Inc., 1 PPG Place, Pittsburgh, Pennsylvania 15272, USA

R. P. MERRETT

> British Telecom Research Laboratories, Martlesham Heath, Ipswich, Suffolk IP5 7RE, UK

Chapter 1

INTRODUCTION TO PLASTICS AND THEIR IMPORTANT PROPERTIES FOR ELECTRONIC APPLICATIONS

MARTIN T. GOOSEY

Dynachem Corporation, Tustin, California, USA

1.1. INTRODUCTION TO PLASTICS AND THEIR PROPERTIES

The word plastic has its origins in the Greek language where 'plastikos' meant to form or shape. Today plastic has a number of meanings but the one which best fits the description of materials detailed in this work is that referring to any of various substances which harden and retain their shape after being moulded or formed by heat and pressure. This description is still rather broad covering a large number of materials including many such as tar, clay and even mud which do not really agree with our current perception of a plastic. In order to qualify the word plastic a more accurate description would be synthetic plastic, since this eliminates the naturally occurring materials that have plastic properties and only includes those derived by chemical means from relatively simple raw materials. Plastics are thus man-made polymers where the word polymer is an abbreviation of the two words poly and monomer and literally means many monomers.

During the nineteenth century much interesting work was performed on polymeric materials beginning with the vulcanisation of rubber in 1839 and also including the development of nitrated cellulose (celluloid) in 1868. It was not until after the turn of the century that the first commercially successful fully synthetic plastic was produced. This was bakelite made from phenol and formaldehyde by Leo

Baekeland in 1909. Before the end of the 1920s a large number of other synthetic polymers had been created, notable examples being the development of polyvinyl chloride in 1927 and urea formaldehyde in 1929. Today there are literally hundreds of synthetic polymers commercially available with ranges of properties making them suitable for applications in many industries including the electrical and electronics industries. For every material commercially available many more have been synthesised, examined and discarded as of no present use for various technical or economic reasons. The year of introduction and typical electronics applications for various plastics are shown in Table 1.1.

Much of the driving force behind the development of many modern plastics materials came from the electrical and electronics industries, largely because of the numerous areas for application in electrical and electronic components. Typical applications for plastics in the electronics and electrical industries are shown in Table 1.2.

The principal reasons for this popularity are that plastics are generally inexpensive, readily shaped, dielectric materials with easily controlled physical, mechanical and electrical properties.

A typical radio manufactured 30 years ago would almost certainly have had a case made from a dark brown 'bakelite' type phenolic resin

Table 1.1
Year of Introduction and Uses for Various Plastics

Year	Material	Application
1909	Phenol-formaldehyde	Equipment cases
1927	Polyvinyl chloride	Wire and cable insulation
1938	Polystyrene	Capacitor dielectrics
1942	Polyethylene	Capacitor dielectrics, wire insulation
1947	Epoxides	Adhesives, potting and moulding compounds
1957	Polypropylene	Capacitor dielectrics, wire insulation
1964	Polyimide	Dielectrics, die attach adhesives
1965	Polyxylylenes	Insulation, conformal coatings
1975	Polymethyl pentene	Microwave components, cable dielectrics
1982	Polyether imide	High reliability chip carriers

Table 1.2
Typical Applications for Plastics in the Electronics and Electrical Industries

Integrated circuits, transistors, diodes and other discrete devices	Transfer moulding compounds, injection moulding compounds, casting, potting, dip and powder coating compounds, die attach adhesives. Photoresists. Junction coatings.
Wires and cables	Sleevings, coatings, varnishes, intermetal dielectrics.
Connectors	Transfer moulding, injection moulding, compression moulding compounds.
Hybrids	Conductive and non-conductive adhesives, sealants, conformal coatings.
Transformers, coils, bushings	Transfer moulding compounds, coatings, potting compounds, coil impregnants, wire insulation.
Printed circuit boards	Laminates, conformal coatings, solder masks, masking tapes, component attachment adhesives, vibration dampers. Photoresists.

and inside this would be components covered in a variety of strong-smelling sticky waxes or bituminous materials. Since then steady improvements in plastics technology have paved the way for a huge selection of materials to be manufactured that are used in electronics applications unimaginable in the 1950s. Today, plastics are used throughout all parts of the electronics industry from the manufacturing processes to the final packaging as well as in the individual microcircuits themselves. Some further applications for plastics in electronic and electrical component manufacture are shown in Table 1.3.

Polymeric materials are used as the basis of photoresists for making metal patterns on printed circuit boards, and also on the individual integrated circuit devices themselves. The circuit boards are made of various glass-fibre filled epoxides and polyesters, and are usually coated with other polymeric materials to prevent tarnishing and to improve solderability. The devices that are mounted on the completed printed circuit boards may have been manufactured using polymeric photoresists or have plastics such as polypropylene and polyester films in capacitors or epoxide resins in integrated circuits and transistors. The completed boards containing various devices are often then coated in another protective polymer. Finally, electrical connections are made

4

Table 1.3

Uses for Plastics in Electronics and Suitable Materials

Use	Suitable materials
Transfer moulding	Epoxides, silicones, phenolics, polyimides.
Injection moulding	Polyethylene, polypropylene, polyphenylene sulphide.
Encapsulation/casting	Epoxides, polyurethanes, silicones.
Adhesives	Epoxides, polyimides, cyanoacrylates, polyesters, polyurethanes.
Coatings	Silicones, fluorocarbons, epoxides, polyxylylenes, polyurethanes, polyimides.
Films	Polyesters, polypropylene, polystyrene, polyimides.
Sealants	Silicones, polysulphides, polyurethanes, epoxide-polyamides, fluoro-silicones.

with wires coated in one of several plastics and the finished circuitry is housed in an attractive plastic case. Plastics can thus be seen to play an important role throughout all parts of the electronics and electrical industries from the initial fabrication of devices to their final packaging.

1.2. CHEMISTRY OF PLASTICS

Polymers are different from the majority of simple inorganic and organic molecules commonly encountered in that they have very large molecules with molecular weights sometimes exceeding 1 million. Even though these molecules are large, they can be defined by the basic repeat unit that forms the polymer chain. Many of the overall properties of a particular type of polymer will be a direct result of the individual atoms and the way they are arranged in the repeat unit.

One of the most important atoms in polymer chemistry and indeed in all organic chemistry is the carbon atom. In polymer chemistry carbon is the atom that forms the main chain in most molecules. Carbon has a valency of four, which simply means that it has four sites at which it can combine with other elements. If a single atom of carbon combines with four atoms of hydrogen the gas methane is formed

$$H-\underset{\underset{H}{|}}{\overset{\overset{H}{|}}{C}}-H$$

Methane (CH_4)

Carbon can combine with many other elements besides hydrogen and can also combine with itself. Examples of carbon atoms combined with themselves and hydrogen atoms are the alkanes, such as ethane and propane, where two and three carbon atoms are joined together, respectively, and the rest of the unoccupied sites are used to combine with hydrogen atoms

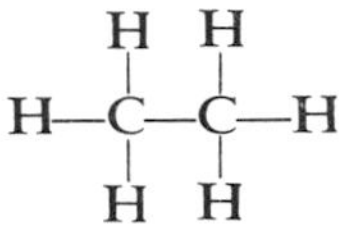

$$
\begin{array}{ccccc}
 & H & H & & \\
 & | & | & & \\
H- & C- & C- & H & \\
 & | & | & & \\
 & H & H & &
\end{array}
\qquad
\begin{array}{ccccccc}
 & H & H & H & & \\
 & | & | & | & & \\
H- & C- & C- & C- & H & \\
 & | & | & | & & \\
 & H & H & H & &
\end{array}
$$

Ethane (C_2H_6) 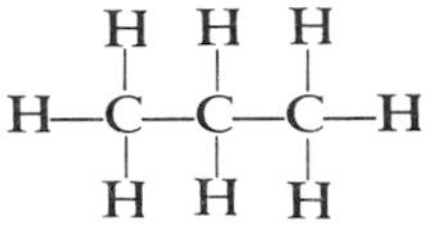   Propane (C_3H_8)

One of the features of carbon which leads to the formation of polymers is its ability to form double bonds with itself. In the molecule ethylene (C_2H_4) there are not enough hydrogen atoms to occupy all of the available bonding sites and so the carbon atoms form a double bond between themselves

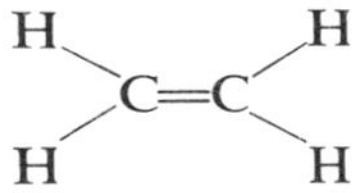

Ethylene (C_2H_4)

These double bonds are very reactive and under the right conditions ethylene molecules can polymerise in a chain reaction to form a long linear molecule of polyethylene

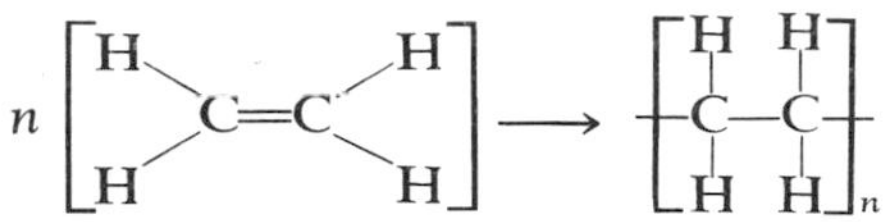

Ethylene   Polyethylene

Carbon also has the capability of forming ring structures containing various numbers of carbon atoms; the simplest molecule containing six carbon atoms in a ring structure is benzene

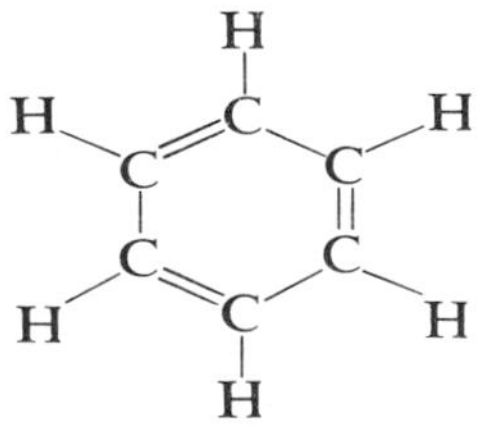

Benzene (C_6H_6)

For simplicity, the benzene molecule is drawn as a hexagon with a circle representing the double bonds between the carbon atoms

If one of the hydrogen atoms of an ethylene molecule is replaced by a benzene ring the compound styrene is formed

Styrene

This can also be polymerised by reacting the carbon–carbon double bond of the ethylene part of the molecule to give polystyrene

$$n \left[\begin{array}{c} CH{=}CH_2 \\ \end{array} \right] \longrightarrow \left[\begin{array}{c} H \quad H \\ C{-}C \\ H \end{array} \right]_n$$

Styrene   Polystyrene

In addition to the substitution of one of the hydrogen atoms of ethylene with a benzene ring there are several other substituents that can be used and these lead to the formation of various well-known polymers

Substituent	Resulting polymer
—H	Polyethylene
—CH$_3$	Polypropylene
—Cl	Polyvinyl chloride
—CH$_2$—CH$_3$	Polybutylene
—O—$\overset{\overset{\displaystyle O}{\|\|}}{C}$—CH$_3$	Polyvinyl acetate
—OH	Polyvinyl alcohol
—C≡N	Polyacrylonitrile

These are known as polymers derived from vinyl monomers and the word vinyl refers to an ethylene molecule with one of its hydrogen atoms replaced with another atom or molecule.

There is no reason why only one of the four hydrogen atoms can be replaced and it is quite normal to replace more than one of these atoms. When two atoms are replaced the resultant polymers are said to be derived from vinylidene monomers, where vinylidene refers to an ethylene molecule with two of its hydrogen atoms replaced with other atoms or molecules. When two hydrogen atoms are replaced they are both from the same carbon atom of the ethylene molecule. Some examples of linear polymers based on vinylidene monomers are shown below with their substituents.

Substituent A	*Substituent B*	*Resulting polymer*
—Cl	—Cl	Polyvinylidene chloride
—F	—F	Polyvinylidene fluoride
—CH$_3$	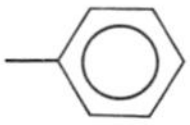 —⬡	Poly α-methyl styrene
—CH$_3$	$-\overset{\overset{\text{O}}{\|\|}}{\text{C}}$—O—CH$_3$	Poly methyl methacrylate

One further group of polymers based on ethylene is possible and these have all four hydrogen atoms of the ethylene molecule replaced. When the substituent is fluorine the resultant polymer is polytetrafluoroethylene; alternatively, the hydrogen atoms may be replaced by three fluorine atoms and one chlorine atom in which case the resultant polymer is polychlorotrifluoroethylene

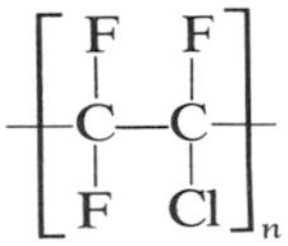

$$\left[\begin{array}{cc} \overset{\text{F}}{\underset{\text{F}}{\mid}} & \overset{\text{F}}{\underset{\text{Cl}}{\mid}} \\ -\text{C}- & -\text{C}- \end{array}\right]_n$$

The above examples have been based upon ethylene and its derivative monomers, all of which have led to the formation of polymer backbones containing chains of linked carbon atoms. Carbon may join with other atoms and molecules such that they can also be incorporated into the polymer backbone. The previous examples have all been formed by the addition polymerisation process in which carbon–carbon double bonds have been changed to single bonds in order that further carbon atoms could join together. In many of the polymers containing different atoms in their backbones polymerisation may have occurred by another means such as condensation polymerisation. These processes are covered further in Chapter 2. Typical examples of linear polymers containing other atoms in addition to carbon in their backbones are shown below

| *Structural unit* | *Polymer* |

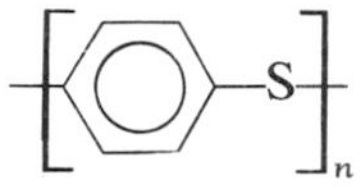

Polyphenylene sulphide

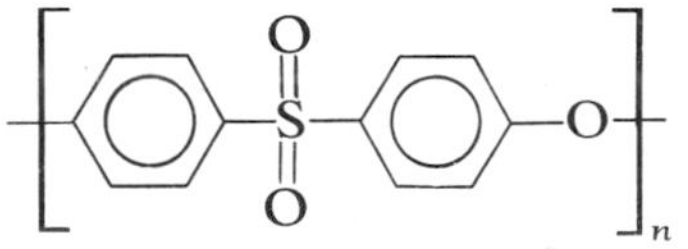

Polyether sulphone

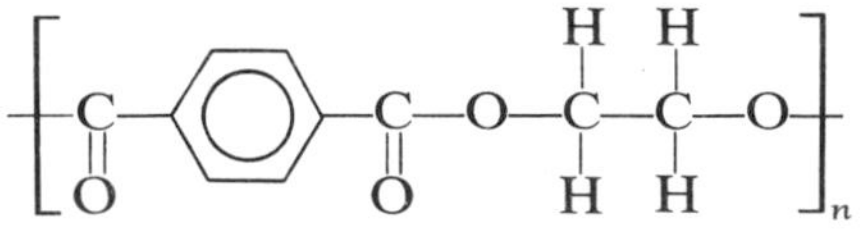

Polyethylene terephthalate

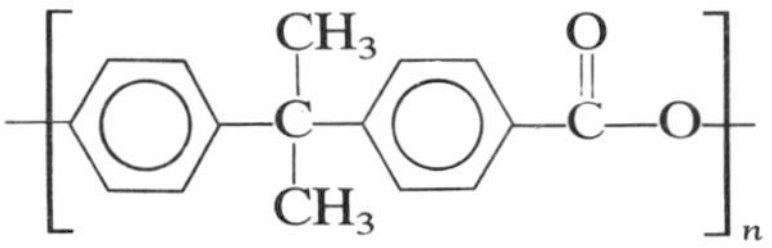

Polycarbonate

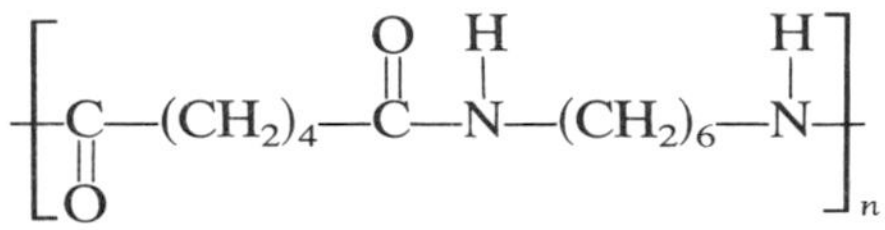

Nylon 66

Although carbon has the ability to form very large long chain molecules it is not completely unique in this characteristic. Nitrogen and oxygen can form long chain molecules with carbon atoms. Sulphur can be either an individual sulphur atom as in polyphenylene sulphide or it may form short chains of sulphur atoms. Perhaps the main examples of long chain polymers not requiring carbon atoms to form their backbones are the silicones. In these materials the main chain is comprised of alternating silicon and oxygen atoms

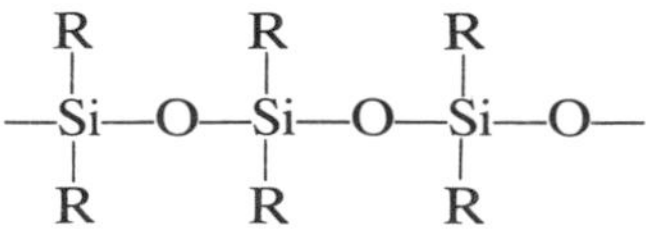

These materials are covered in detail in Chapter 3.

1.3. IMPORTANT ELECTRICAL PROPERTIES
OF POLYMERS

One of the main reasons why plastics have become so widely used in electronics and electrical applications is that they are good dielectric materials with readily controllable electrical properties. However, the fact that plastics are good insulators does not mean that they are inert in an electrical field and the intrinsic properties of a material can usually be related to performance under specific test conditions. The electrical properties that have to be considered when selecting a material are the dielectric constant, the volume resistivity, surface resistivity, dissipation, power and loss factors, arc resistance and dielectric strength. When selecting a plastic for a specific application the variation of these properties with changes in the likely service environment has to be taken into account since many of them are temperature, frequency, voltage and environmentally variable.

Most plastics serve as electrical insulators and may be required to operate in conditions of varying voltage with both direct current and alternating current at frequencies up to and above 10^5 MHz. The voltage the plastic is exposed to may be as little as a few microvolts in communications equipment or up to several million volts in power distribution systems. Similarly, currents can vary from a few picoamps to many thousands of amps. All of these variations may be encountered over a wide range of temperatures, humidities and operating environments. The selection of a material capable of operating under such variable conditions is obviously important and thus an understanding of the electrical properties of plastics and how they vary under these conditions is essential.

Some of the more important electrical properties are now discussed.

1.3.1. Resistance and Resistivity

Most plastics, unless purposely modified, are excellent insulators having very high resistivities and conduct little electrical current. The insulation resistance of a material is the ratio of the applied voltage to the total current passing between two electrodes contacting the material. This resistance is measured in ohms and is directly proportional to the length and inversely proportional to the area of the specimen according to the equation

$$R = \frac{\rho l}{A}$$

where R = insulation resistance (Ω), l = length (m), A = area (m^2), ρ = specific resistivity (Ω m).

In order to compare different materials it is usually more convenient to compare their resistivities because these values reduce resistance measurements to a common denominator. Volume resistivity is the ohmic resistance of a cube of dielectric material with sides of length 1 cm or 1 m, and is expressed in Ω cm or Ω m, respectively.

All materials may be classified by their resistivities and they range from the metals which are extremely good conductors to the plastics which are very good insulators. Between these two extremes are the semiconductors such as germanium, silicon and gallium arsenide. Most materials can thus be conveniently placed into one of three groups: those that are good conductors, those that are good insulators and those in between which may be semiconductors. The resistivities of a number of materials are given in Table 1.4. Plastics have resistivities of 10^{12} Ω cm or greater and some fluorinated polymers have resistivities in excess of 10^{18} Ω cm.

Various factors influence the resistivity of a plastic and these have to be taken into account when formulating or selecting materials for use in electronics. The chemical structure of the polymer itself will play an important role in determining the resistivity, as will the concentration of any impurities remaining in the material such as by-products of manufacture. The addition of flexibilising and plasticising agents is also known to reduce the resistivity of plastics and the degree of cure or crosslinking density of thermosetting materials must also be considered.

Resistivity varies with temperature and, in an opposite manner to metals, the resistivity of polymers decreases with temperature. There is thus the possibility that a material that has adequate insulative properties at room temperature may be too conducting at elevated temperatures. The temperature dependence of resistivity is given by the equation

$$\rho = \rho_0 \exp (\Delta E / 2kT)$$

where ρ = resistivity at temperature T, ρ_0 = limiting low temperature resistivity, k = Boltzmanns constant, ΔE = energy, T = absolute temperature.

Other factors influencing the resistivity of a plastic include voltage, frequency, pressure and the conditioning of the sample. In particular, environmental conditions are very important since relative humidity

Table 1.4

Comparative Electrical Resistivities of Materials

Resistivity (Ω cm)	Material	Classification
10^{18}	Polytetrafluoroethylene	
10^{17}	Polystyrene	
10^{16}	Polyethylene, polypropylene	
10^{15}	Amine-cured epoxides	
10^{14}	Diamond	
10^{13}	Polyamides	Insulators
10^{12}		
10^{11}		
10^{10}	Glass	
10^{9}		
10^{8}	Silver bromide	
10^{7}		
10^{6}		
10^{5}		
10^{4}	Silicon	Semiconductors
10^{3}		
10^{2}	Germanium	
10^{1}		
10^{-1}		
10^{-2}	Quinolinium salts	
10^{-3}		
10^{-4}	Bismuth, mercury, graphite	Conductors
10^{-5}		
10^{-6}	Silver, copper	
10^{-7}		

changes alter the resistivity. If a plastic material absorbs a significant amount of moisture, the resistivity may drop to an unacceptably low value. Whilst this increase in conductivity is generally due to the direct presence of sorbed moisture within the polymer matrix and is reversible if the material is dried, some plastics may actually be degraded by the presence of moisture and high potential gradients. One example of this behaviour is the polyester materials which, when exposed to high humidities in the presence of an electric field, may be hydrolysed to the acid and alcohol constituents from which they were made. In the presence of moisture these then form a conductive solution which leads to corrosion or short-circuiting.

The surface resistivities of plastics are also important because moisture and contaminants affect surface resistivity more than volume resistivity. Large mouldings may take several weeks or months to register any changes in volume resistivity due to varying humidity conditions, but under the same humidity conditions or in dirty environments the surface resistivity may change almost immediately. The surface resistivity of a material is the ratio of the potential gradient parallel to the current along its surface to the current per unit width of surface and is numerically equal to the surface resistance between two electrodes forming opposite sides of a square. The size of the square is immaterial and the units are Ω/square.

For some special applications conductivity may be more important than resistivity, as is the case with conductive adhesives. The conductance of a polymer is the reciprocal of its resistance and likewise its conductivity is the reciprocal of its resistivity. The units of specific conductivity are mhos per centimetre or Siemens per metre. As polymers are normally good insulators, conductivity is imparted by the incorporation of finely divided metals such as silver, gold and copper. Conductive polymers find important uses as electromagnetic and radio frequency shielding, as well as die attach materials for semiconductors.

1.3.2. Dielectric Constant (permittivity)

The dielectric constant is the ratio of the capacitance of a capacitor having a given material as its dielectric to the capacitance of the same capacitor having a vacuum as the dielectric

$$K = \frac{C_m}{C_v}$$

where K = dielectric constant, C_m = capacitance with material m as the dielectric and C_v = capacitance with a vacuum as the dielectric.

The dielectric constants of materials vary from just above 1 to up to 1000 for some inorganic materials, and typical examples are shown in Table 1.5.

The dielectric constants of materials are due to their electronic polarisability and compounds with polar groups in their molecular structure have large dielectric constants because their dipoles are able to orient in an electric field and have large dipole moments. Polar polymers absorb moisture from the atmosphere and for these reasons non-polar polymers are normally preferred over polar polymers for insulative applications. For general insulating or protective coatings dielectric constants up to 8 may be acceptable but the maximum

Table 1.5
Dielectric Constants of Various Materials at 25°C

Material	Dielectric constant
Air	1·00053
Polystyrene	2·45–2·65
Epoxides	2·8–4·6
Phenolics	5·0–6·5
Polypropylene	2·2–2·3
Nitrobenzene	34·9
Water	79·5

dielectric constant specified by MIL-I-16923 for materials used in electronic and electrical assemblies is 4·5 at 1 KHz.

Dielectric constants of polymers vary with frequency and can either increase or decrease depending upon molecular structure, and the presence of any additives or fillers. In order to be of use in electrical and electronic applications a plastic should have reasonably constant dielectric properties over a broad range of temperatures, frequencies and humidities. It is for these reasons that materials such as polyethylene, polystyrene and the fluorocarbons find such extensive usage as electrical insulators.

1.3.3. Dissipation, Power and Loss Factors

The dissipation factor (D) is the ratio of the current of the resistive component (I_r) to the current of the capacitive component (I_c) and is also equal to the tangent of the dielectric loss angle δ, i.e.

$$D = \frac{I_r}{I_c} = \tan \delta$$

Typical values of the dissipation factor vary from 0·1 to less than 0·0005.

The power factor (PF) is the ratio of power dissipated to the product of the effective power output (volts × amps) and is a measure of the dielectric loss in the insulating plastic (which is effectively acting as a capacitor). The power factor and dissipation factor are related by the equation

$$D = \frac{PF}{\sqrt{[1 - (PF)^2]}}$$

Power factors for most plastics are fairly low and so the $(PF)^2$ term in the equation can be ignored, making the power factor approximately equal to the dissipation factor. These two terms are often used interchangeably for dielectric materials.

The loss factor is a measure of the energy absorption of a material and is given by the product of the dielectric constant and the power factor, i.e.

$$\text{Loss factor} \simeq K \tan \delta \simeq KD$$

It is desirable for all of the above parameters to have as low values as possible and this is particularly important in the case of high-speed, high-frequency devices and circuitry working at frequencies up to 10^5 MHz. Low values are indicative of minimum conversion of electrical energy to heat and little overall power loss in the system. These factors are again generally functions of frequency, temperature, humidity and the purity of the polymer.

1.3.4. Arc Resistance

The arc resistance of a material is the time (in seconds) that its surface may be exposed to an arc before electrical breakdown occurs. A high-voltage, low-current arc is used to simulate the conditions likely to be found in service. A standard test procedure used in the United States is described in ASTM-D-495 and a number of arc testers are commercially available that comply with this specification. The specimen to be examined is placed between the electrodes and an arc generated at specified intervals and currents.

Failure due to arcing can occur in three different ways:

(1) By the formation of a thin conductive line between the electrodes known as tracking.
(2) By the formation of a conductive path of carbonaceous material due to surface heating.
(3) By self-ignition of the material without the formation of any visible conductive path.

The arc resistance of a plastic is very dependent upon its molecular structure and the presence and amount of any additives in the formulation. Fillers have a significant effect upon the arc resistance and materials such as alumina and silica can give considerable improvements. Arc resistance is also very dependent on the condition of the surface and is easily reduced by moisture, salts or even grease from skin contact.

1.3.5. Dielectric Strength and Breakdown Voltage

The dielectric strength of a polymer is a measure of its ability to withstand voltage without breakdown or the passage of considerable amounts of current. It is defined as the minimum voltage at or below which no breakdown occurs. Rather similarly, the breakdown voltage is the voltage above which actual failure occurs and the two items are often used interchangeably.

The dielectric strength of a material is, like the properties reported above, very dependent upon molecular structure, the level of impurities and the relative humidity to which the sample has been exposed. Consequently the dielectric strength of a material should be measured under carefully specified conditions. There are several standardised procedures for measuring dielectric strength and in the United States ASTM-D-149 is commonly used.

1.4. IMPORTANT MECHANICAL, CHEMICAL AND PHYSICAL PROPERTIES OF POLYMERS

In addition to having excellent electrical properties for electronic and electrical applications, polymers also have desirable mechanical, chemical and physical properties that enhance their capabilities in these fields. Plastics are available with very high purity and can be easily moulded to form rigid protective packages for delicate microcircuits, whilst others are available that are extremely flexible making them useful as wire coatings and sleevings. These properties are determined largely by the chemical composition of the given material but the way in which a plastic is manufactured and used sometimes also has a bearing on the final properties. In this section some of the more important mechanical, chemical and physical properties of polymers are discussed.

1.4.1. Mechanical Properties

Unlike many metals polymers are not perfectly elastic, possessing characteristics of both elastic and viscous materials. Consequently they are often regarded as being viscoelastic. In order to select the best material for a given application it is usually important to know how its properties vary with time, temperature and environmental changes. A number of the more important physical capabilities of a polymer can be defined by examining the following properties: elasticity, stiffness, strength, toughness and resilience.

The elasticity of a polymer is its ability to carry stress without suffering a permanent deformation. The stiffness of a material is a measure of its ability to carry a stress without changing its dimensions, and strength determines its ability to carry a dead load. Toughness represents the ability to absorb energy and undergo permanent changes without rupturing whilst resilience is a measure of the work required to deform a material to its elastic limit and determines the energy that it can absorb without undergoing a permanent deformation.

All of these properties can be determined from the stress/strain characteristics of a plastic by plotting its stress/strain curve. Stress is defined as the intensity (measured per unit area) of the internal distributed forces, or components of force, which resist a change in the form of a body. Strain is the change per unit length in a linear dimension of a body accompanying a stress and is measured in metres per metre or in percent. When under tension or compression, strain is measured along the dimension under consideration. Many plastics have stress/strain characteristics that can be represented by a curve with the general shape shown in Fig. 1.1.

The slope of the curve where stress and strain are almost proportional represents a measure of the material's stiffness. In this region the ratio of stress to strain is known as the modulus of elasticity or Young's modulus. At the first bend in the curve (1) the stress value represents the yield point and is a measure of the strength of the material and its resistance to permanent deformation. Before the yield point is reached

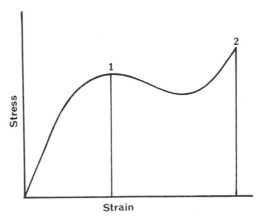

Fig. 1.1. A typical stress/strain curve for a polymer.

strains are largely recoverable and are a measure of elastic deformation. From the yield point (1) to the break point (2) the strains are no longer immediately recoverable and they determine the degree of plastic deformation. The area under the curve represents the work required to fracture the polymer sample and is an approximate measure of its toughness. The resilience of a material can be determined from the area under the curve up to the yield point. The various shapes of stress/strain curve allow polymers to be classified according to five main types of behaviour:

(1) Soft and weak polymers having low moduli of elasticity, yield points and low elongations at break.
(2) Hard and brittle polymers having high moduli of elasticity, no well-defined yield points and low strain at break.
(3) Soft, tough polymers having low moduli of elasticity, low yield points and high strain at break.
(4) Hard, strong polymers having high moduli of elasticity and yield points with moderate stress at break.
(5) Hard, tough polymers having high moduli of elasticity and yield points and also high stress and strain at break.

In addition to mechanical failure under conditions of high stress, failure can also occur at relatively low stress levels under some conditions and this must again be taken into account when selecting materials for a specific application. The failure of a polymer at relatively low stress levels can occur under the following circumstances:

(1) Fatigue
(2) Stress relaxation
(3) Creep
(4) Chemical exposure

Fatigue is the term used to describe the behaviour of a plastic under repeated cycles of stress or strain which cause a deterioration of the material and result in progressive fracture. Fatigue is dependent upon frequency, amplitude, temperature, stress and mode of stressing.

Stress relaxation occurs in a plastic when it has been deformed to a predetermined shape and then restrained in that shape so that total deformation neither increases nor decreases. Under these conditions the phenomenon of stress relaxation can then occur and the initial stress in the plastic gradually decreases. It can be defined as the time-dependent change in stress which results from the application of a constant total strain to the material at constant temperature.

Creep is defined as the time-dependent part of the strain which results from the application of a constant stress to a plastic. By comparing the definitions of creep and stress relaxation it can be seen that the two phenomena are closely related. Creep occurs when a constant stress is applied for an extended period whilst stress relaxation occurs during long-term exposure to a constant strain.

The loss of mechanical properties during exposure of a plastic to certain environments may also be important in electrical and electronic applications. Whilst some chemicals may have little effect upon the strength of an unstressed part there may be a totally different behaviour with the same part under stressed conditions. Contact with water, acids, alkalis and solvents can cause a number of undesirable effects in a plastic such as swelling, shrinkage, distortion, crazing, cracking, softening, weight loss or gain, and loss of strength or stiffness. All of these factors must be taken into account when selecting a plastic for use in electrical and electronic components, particularly where mechanical and physical properties are likely to be important.

A number of other mechanical properties are also often considered important. The hardness, abrasion resistance and machinability may need to be considered where some reworking of a compound is needed or where machining into a particular shape is required. Depending upon the final usage of the material the tensile, shear and compressive strengths of a plastic can be important. Such properties would need to be considered where the plastic was used as a structural component or where it is likely to be exposed to forces that could cause mechanical failure. The flexibility of a polymer and, more particularly, maintenance of flexibility over a likely operating temperature range need to be characterised since wire coatings must stay flexible at low temperatures and junction coatings used to protect semiconductors must not become rigid if equipment they are used in happens to be exposed to low temperatures. Shrinkage can be important when attempting to mould plastic parts to strict tolerances since thermosetting materials experience shrinkage during moulding because of the curing reaction which takes place. Another property that is important with plastic moulding compounds is viscosity. This must be controlled so that moulds fill properly yet the resinous components do not bleed out of the mould, and, in the case of wire-bonded semiconductors, so that wire bonds are not swept off contacts causing shorting or open circuits.

1.4.2. Chemical Properties

The resistance of a polymer to various chemicals is largely determined by its molecular composition and varies widely from type to type. Some plastics such as polytetrafluoroethylene are inert in even the most aggressive chemicals and solvents whilst others are easily decomposed in the mildest solvents or are soluble in water. Although exceptions are frequently discovered it is quite possible to make predictions of a material's likely behaviour in various environments by studying its molecular structure. Polymers that have a backbone made up entirely of carbon–carbon bonds with hydrophobic groups attached usually have good oxidation resistance and low moisture absorption. Low moisture sorption obviously helps to minimise attack by aggressive aqueous solutions. If a degree of double bonding (unsaturation) is introduced into the chain the oxidation resistance will tend to decrease although the water absorption characteristics may change only slightly. A typical example is butyl rubber that contains few double bonds. This is much more resistant to oxidation than natural rubber which has a larger number of double bonds in the chain.

Where the polymer backbone contains groups such as the ester or amide linkages found in polyesters and polyamides the degree of moisture sorption is usually high and they are readily attacked by chemicals with hydrolysing capabilities. They are thus rather unstable in both moderately strong acids and alkalis. The absolute structure of the material concerned will play a part in determining the actual degree of hydrolysis that occurs and the moisture sorption characteristics will also be important. If the molecular structure dictates a low moisture permeability, transport of aggressive solutions into the bulk of the material may be retarded with attack predominantly limited to the materials surface.

The introduction of hydrophilic groups onto the main chain of the polymer results in the enhanced sensitivity of the material to both moisture and aqueous solutions. The classic example of such a material is polyvinyl alcohol which is soluble in both water and some other polar solvents. Interestingly, however, polyvinyl alcohol is not affected by hydrophobic solvents such as the hydrocarbons. The presence of an hydroxyl group on every second carbon atom in the main chain of polyvinyl alcohol not only provides high polarity to the molecule but also allows it to exhibit strong adhesion to certain substrates. In general the introduction of polar groups both in the main chain and in the side groups leads to very high surface energies and surface tensions resulting in good adhesive properties. The amide linkage in polyamides, whilst being largely responsible for their moisture sorption, also leads to powerful hydrogen bonding and their good adhesive properties.

The molecular weight of a polymer molecule also has a significant effect in determining the chemical properties. A typical example is a polymers' solubility in a solvent which is usually inversely proportional to its molecular weight. The swelling, absorption and permeability of polymers to gases and liquids is also related to molecular weight although the relationship is not clearly defined and it has been stated that molecular weight does not affect diffusion and permeation by liquids and gases. The general chemical reactivity of polymers is known to decrease with increasing molecular weight. This is thought to be due to the reactivity of the end groups of the polymer molecule which are usually the most reactive parts of the whole molecule. Since the number of end groups is inversely proportional to the number average molecular weight it is expected that polymer reactivity will decrease with increasing molecular weight. As ageing processes are chemical degradation reactions that occur in polymers, their resistance to this type of deterioration should also be related to molecular weight. The ageing of polymers by oxygen, ozone, moisture and ultra-violet light is found to decrease with increasing molecular weight and again this is because the number of reactive end groups is reduced. The lower mobility of the higher molecular weight molecules also reduces reactivity and degradation of very high molecular weight polymers may only degrade them to medium molecular weight polymers with almost equivalent properties.

The presence of impurities or monomeric additives in a polymer also has an important bearing upon its final properties and the applications it will be suitable for. No organic chemical reactions give absolutely 100% yields of the desired products and polymerisation reactions are no exception. In addition to by-products, residual monomers may be present which can degrade the mechanical and thermal properties of the plastic. Residual solvents can cause softening and degradation of electrical properties in polymers, particularly where coatings from solution are being applied or where polymerisation is carried out in solution. Moisture and other small molecules can be generated during condensation polymerisations or they can be absorbed from unsuitable storage environments. The presence of moisture, whilst not only causing degraded mechanical and electrical properties, can lead to difficulties in moulding a polymer and may even be the source of hydrolytic instability. Contamination can also occur in the form of ions which are residual impurities from the synthesis of the polymer's starting materials. Chloride and sodium ions are found in epoxide resins as by-products of the synthesis and these can lead to malfunctioning of epoxide encapsulated semiconductors and early corrosion of certain conductors such as aluminium. In addition, ions can lead to reductions in the insulative properties of a polymer. Catalysts used to promote a polymerisation reaction may leave residues that have a deleterious effect upon thermal stability and electrical properties of a polymer. Effective post-cures are required with epoxide moulding compounds in order to decompose these catalyst residues and to improve the bulk resistivity of the cured materials.

In order to evaluate a material for its chemical stability it is necessary to test it under conditions which as nearly as possible replicate those likely to be encountered in service. One fairly simple test that can be applied to most plastic materials as part of an initial screening process, involves immersing moulded discs or bars for seven days in a variety of liquids at room temperature. This test uses a number of acids, alkalis and salt solutions of varying concentrations and a series of solvents with a range of polarities. At the end of the test the samples are examined for changes in dimensions, weight, colour and other physical properties.

Before selecting a polymer for an electrical or electronic application its chemical properties should be considered just as the electrical, mechanical and physical properties are taken into account. The final operating environment and conditions of service will often dictate the type of material to be used and the required purity.

1.4.3. Physical Properties
A number of other important properties have to be considered when selecting a plastic for electronic and electrical usage. There may be restrictions on the amount of material that can be used or perhaps the plastic must have a certain degree of thermal conductivity to impart heat dissipation to a circuit. Some of these other physical properties are now briefly discussed.

Plastics have distinct advantages over the majority of other materials in terms of weight of material used. Most plastics have densities that are considerably lower than the metals they often replace in structural applications with values normally between 0·9 and 2·0 (a good number lie between 1·0 and 1·4). Density considerations become important in applications where overall weight must be minimised such as in aerospace electronics. Since the approval of a plastic used is normally based on volume rather than weight and yet the plastic is sold by weight, a reduction in density of the material will also allow more parts to be made per unit weight and this factor should be considered where material costs must be carefully controlled.

There are several thermal properties of a polymer that are considered important, such as its thermal expansion coefficient, thermal conductivity, thermal stability, melting point and glass transition temperature. The thermal expansion coefficient of a plastic needs to be taken into account when it is in intimate contact with another material with a different thermal expansion coefficient. Plastics have much higher thermal expansion coefficients than most other materials especially metals, ceramics and the silicon substrates found in electrical and electronic components. The thermal expansion coefficient depends partly upon molecular weight and this is largely because the polymer end groups are one of the major contributing factors. Increasing the effective molecular weight by crosslinking also assists in reducing the expansion coefficient of a plastic. Where lower thermal expansion coefficients are required, fillers are often added to reduce them as is the case with most moulding compounds.

The thermal conductivities of polymers are much lower than those of other materials and because of this they often find use as thermal insulators. The thermal conductivity coefficient is the time rate of heat flow under steady-state conditions through a unit area and unit thickness per unit temperature gradient. Plastics used with electrical and electronic assemblies usually require high thermal conductivities because of the need to dissipate heat from circuitry. For many applications the plastics are formulated with high thermal conductivity fillers such as crystalline silica or alumina. One of the best thermally conductive electrically insulating minerals is beryllia (BeO) although it has to be used with care because of toxicity problems. Where possible the thermal conductivity can also be improved by using as thin a coating as possible.

The thermal stability or heat resistance of polymers is obviously of particular importance when products have to withstand exposure to heat under a wide variety of service conditions. As plastics are organic materials they are susceptible to the reactions of these materials such as thermal degradation, oxidation and hydrolysis. Thermal stability is another parameter that is directly related to molecular weight and the polymer's molecular structure. Much of the instability at high temperatures is due to the nature of the end groups of the polymer molecules, which are sites for reactivity. The thermal stability of a polymer can be enhanced by reducing the number of reactive impurities that can take part in thermal degradation. Crosslinking a polymer increases its molecular weight and decreases its volatility leading to increased thermal stability. Molecular structure also plays a significant part in determining thermal stability and the presence of aromatic and highly electronegative elements tends to enhance both the thermal and the oxidative stability. The hydrogen atoms of polyethylene can be replaced with electronegative fluorine atoms to give the high thermal stability polymer polytetrafluoroethylene.

The melting point and/or the glass transition temperature (T_g) of a plastic will also play an important part in the choice of materials for an application. Both of these thermal characteristics play a large part in determining many of the mechanical properties of a plastic. The glass transition temperature of a material represents the point above which the plastic behaves as a viscous liquid or a rubber. Below the glass transition temperature the plastic behaves as a hard, brittle, glass-like compound. Silicone rubbers are examples of materials that have T_g values below room temperature and perspex is an example of a material that has a T_g above room temperature. The T_g of a material is determined by its molecular structure with factors such as flexibility, symmetry and the bulkiness of the molecules playing a major part. The subject of glass transition temperatures is discussed further in

Table 1.6

Glass Transition Temperatures (T_g) and Melting Points (T_M) of some Common Plastics

Plastic	$T_g(°C)$	$T_M(°C)$
Polyethylene	−120	137
Polypropylene	−18	176
Polystyrene	100	240
Polyethylene terephthalate	69	~267
Polymethyl methacrylate	105	160
Polyvinyl chloride	87	212

Chapter 2. Obviously the melting point of a polymer also determines the acceptable temperature range over which it can be used and for thermoplastic materials will limit the maximum service temperature of, for instance, coated wires. Thermoplastic and thermosetting materials are also covered in more detail in Chapter 2. The glass transition temperatures and melting points of several thermoplastic materials are given in Table 1.6.

Chapter 2

FUNDAMENTAL PROPERTIES OF POLYMERS FOR ELECTRONIC APPLICATIONS

J. COMYN

School of Chemistry, Leicester Polytechnic, Leicester, UK

Polymers are widely used as plastics, rubbers, fibres, adhesives, paints and sealants and their use has extended into the field of electronics, where they are used as insulators and encapsulants. In this context they should be good electrical insulators, be free from dielectric losses over the frequency range employed and have a high dielectric strength. Further they should be able to protect electronic components from their surroundings and this in turn depends on them having low permeability to atmospheric vapours and gases, and having good adhesion to the substrate. It is thus the aim of this chapter to explain the chemical and physical structures of polymers and to show how these control their electrical, barrier and adhesive properties.

A full treatment of such a subject would take much more space than is permitted here, and in consequence this chapter is of an introductory nature. However, reference is made throughout to fuller treatments, and the author has chosen to cite books and reviews in preference to original papers, as this both keeps the bibliography to a reasonable size and is probably of more use to the reader. A more detailed treatment of plastics is given in an excellent book by Brydson[1] and values of physical constants can be found in the *Polymer Handbook*.[2] Most numerical values in this chapter are taken from one of these sources.

J. Comyn

2.1. WHAT ARE POLYMERS AND HOW ARE THEY MADE?

Most polymers are organic materials (i.e., carbon compounds) with a structural unit which is repeated a great many times. If for the sake of simplicity we think of the repeat unit as a bead, then a polymer can be made by stringing together a number of beads. Such a polymer is a linear polymer, and can be distinguished from polymers which are branched or crosslinked (see Fig. 2.1).

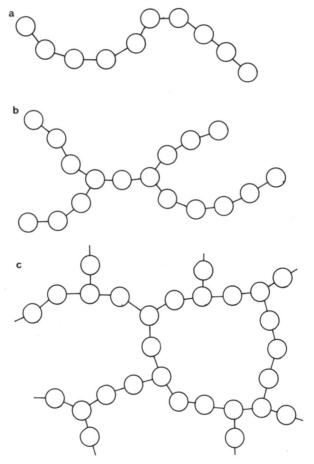

Fig. 2.1. (a) Linear, (b) branched and (c) crosslinked polymers.

It is not easy to distinguish between linear and branched polymers because they have very similar properties, but crosslinked polymers are easily identified because they are insoluble and infusible. The distinction between these two groups of polymers is formally based on their thermal properties. Linear and branched polymers are *thermoplastics* in that repeatedly they may be made to soften and flow by heating. Here flow occurs by polymer molecules sliding against one another when they are hot. Crosslinked polymers are termed *thermosets* and in their final state the crosslinks prevent sliding and flow.

Actually the chemical reactions which cause crosslinking occur when the polymer is moulded, so a convenient way of thinking about the two groups is that thermoplastics can be melted down and used again but thermosets cannot (Fig. 2.2). A general effect of this difference is that thermosets can be used at higher temperatures than thermoplastics; however, at sufficiently high temperatures both types of polymer will chemically degrade.

A few examples will probably be of value at this stage; the repeat units of the polymers mentioned are given in Table 2.1. Polyethylene is widely familiar from its use in plastic bags. It is commercially available as a linear polymer (known as high density polyethylene) and also as a branched polymer (known as low density polyethylene). It is actually possible to crosslink polyethylene by exposure to γ rays, but this is of no commercial importance. Polystyrene is used to make cheap pen barrels and plastic model kits, and polyvinyl chloride is used as a wire insulator and in plastic pipes and clothing. Although polytetra-fluoroethylene is a linear polymer, its very high melt viscosity prevents it from being processed by the usual methods applicable to thermo-plastics. Other common thermoplastics include nylon 6, polymethyl

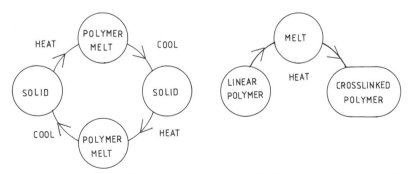

Fig. 2.2. The effect of heat on thermoplastics (left) and thermosets (right).

J. Comyn

Table 2.1.
Glass Transition Temperatures and Melting Points of Some Linear Polymers

Polymer	Repeat unit	$T_g(°C)$	$T_m(°C)$
High density polyethylene (HDPE)	—CH_2—CH_2—	$(-125)^a$	137
Polypropylene (isotactic) (PP)	—CH_2—CH— $\|$ CH_3	-8	171
Polymethylmethacrylate (PMMA)	CH_3 $\|$ —CH_2—C— $\|$ $COOCH_3$	105	Amorphous
Polyvinyl chloride (PVC)	—CH_2—CH— $\|$ Cl	81	^{b}
Polystyrene (PS)	—CH_2—CH— $\|$ (phenyl)	100	Amorphous

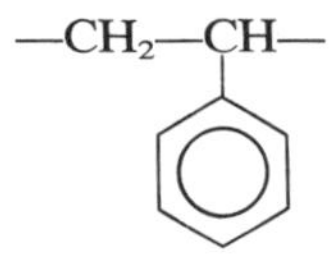

Polymer	Structure		
Polyvinyl acetate (PVA)	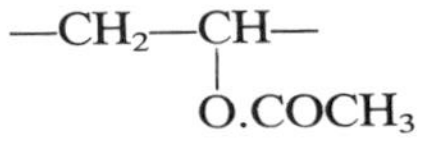 $-CH_2-CH-$ $\quad O.COCH_3$	28	Amorphous
Polyacrylonitrile (PAN)	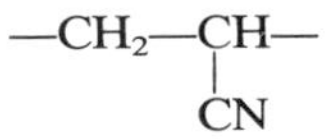 $-CH_2-CH-$ $\quad CN$	105	317[c]
Polyethylene terephthalate (PETP)	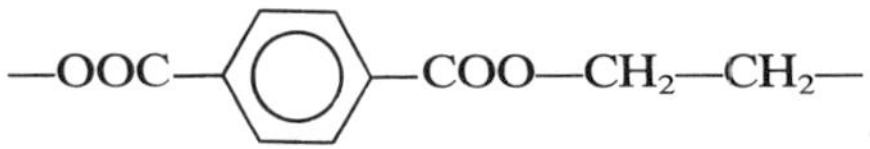 $-OOC-C_6H_4-COO-CH_2-CH_2-$	67[d] 81[e]	250–265
Polydimethylsiloxane	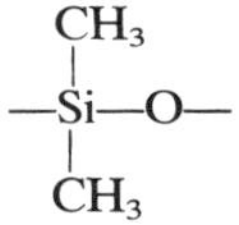 CH_3 $-Si-O-$ CH_3	−127	−29
Nylon 6	$-CH_2CH_2CH_2CH_2CH_2CONH-$	40–52	200–220
Nylon 66	$-NH(CH_2)_6NHCO(CH_2)_4CO-$	50	250–260
Polytetrafluoroethylene (PTFE)	$-CF_2-CF_2-$	−147	342

[a] The T_g of polyethylene is the subject of some controversy.
[b] Slight syndiotacticity gives a low degree of crystallinity.
[c] This atactic polymer has some order which resembles crystallinity.
[d] Amorphous material.
[e] Semicrystalline material.

methacrylate and polypropylene. It is more useful to examine the repeating units of some thermosets as the subject of polymerisation is presented.

Polymers with more than one type of repeating unit are referred to as copolymers, and they are important in that they have properties different from either homopolymer. One example of a copolymer is polyethylene-vinylacetate which is widely used as a hot melt adhesive. Another is acrylic fibre where a small amount of comonomer added to acrylonitrile makes it easier to spin and to dye. In both these examples the monomers are mixed together and then polymerised by a free radical process; the resulting copolymer has a random arrangement of the two units along the chain. Using more specific methods of polymerisation it is possible to prepare block and graft copolymers.

```
    —AABABAAAABAB—              —AAAAAAABBBBBB—
        Random copolymer                  Block copolymer
                                    B
                                    B
                                    B
                                    B
        —AAAAAAA...........AAAA...........AAAA—
            B                               B
            B                               B
            B                               B
            B                               B
            B                               B
                        Graft copolymer
```

With the exception of nylon 6, all the polymers mentioned in the above paragraph are made from a monomer containing a double bond. The principle of polymerisation is quite simple, all we need to do is open up the double bond and then join the pieces together, but the actual mechanism of polymerisation is more complex than this. Polymerisation only takes place when an initiator or catalyst is added to the monomer. The initiator provides free radicals, anions or cations, which are the active centres for polymerisation. Benzoyl peroxide is a commonly used free radical initiator that decomposes upon heating into free radicals which then react with monomer (here vinyl chloride is our example).

Benzoyl peroxide

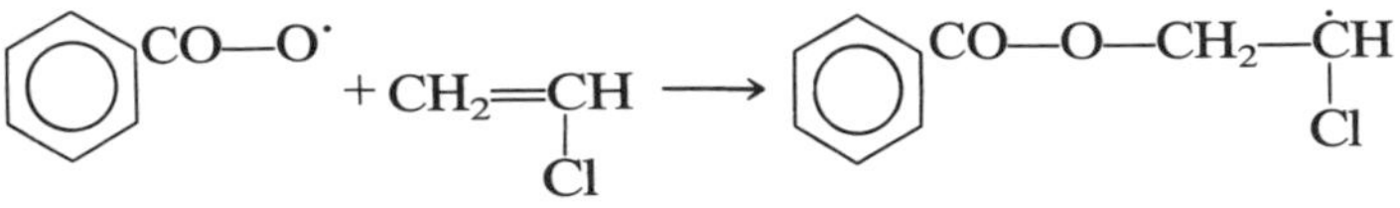

Initiation

These two reactions taken together are known as initiation, and they produce a species with a terminal monomer unit containing an active centre. A single chemical bond consists of a pair of electrons, and in the decomposition of a peroxide the $-O-O-$ bond breaks with one electron remaining with each oxygen atom. Such electrons are unpaired and the species bearing it is known as a free radical, and is represented by a dot.

Further monomer molecules are now added to the free radical in the process which is known as propagation. This reaction occurs repeatedly to produce a molecule containing a linear sequence of monomer units.

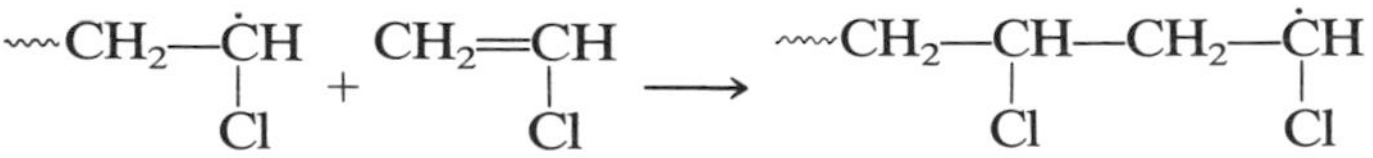

Propagation

The radicals are removed by the process of termination; this may be by either recombination or disproportionation.

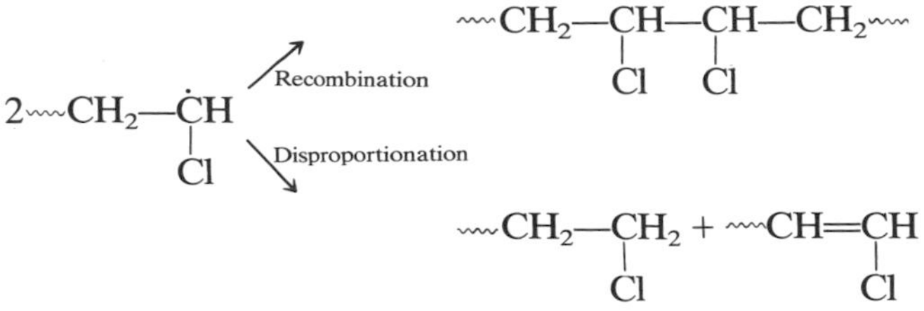

Termination

Thus addition polymerisation consists of the three processes of initiation, propagation and termination which can be compared with the birth, growth and death of a living organism. However, there is a difference in that polymer chains keep growing during all of their lifetime, so a long life leads to a long polymer chain. Free radical addition polymerisation is used in the commercial production of many common polymers including low density polyethylene, polyvinyl chloride, polystyrene and polymethylmethacrylate.

The same principle of opening up a molecule and then joining the pieces together also applies to the polymerisation of ring compounds. Caprolactam is the ring compound from which nylon 6 is made and polymerisation of the epoxide ring is exploited in epoxide adhesives and encapsulants, but the active centre in both these cases is an ion rather than a free radical.

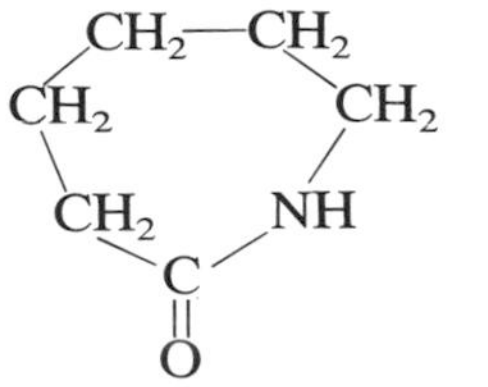

$$\begin{array}{c} CH_2\!-\!CH_2 \\ CH_2 \qquad\quad CH_2 \\ CH_2 \qquad NH \\ C \\ \| \\ O \end{array}$$

Caprolactam

$$\begin{array}{c} O \\ CH_2\!-\!CH\!- \end{array}$$

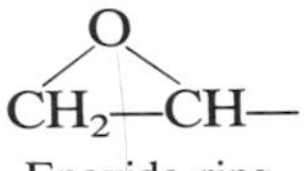

Epoxide ring

The process whereby ring or double-bond compounds give polymers is known as addition polymerisation. Another group of compounds which can polymerise are those containing at least two reactive groups. An example of such a condensation polymerisation is the reaction between 1,6-diaminohexane and adipic acid to produce nylon 66. Here the first compound has two amine groups and the second two carboxylic acid groups. There is a spontaneous reaction between amine and carboxylic groups to produce an amide and water thus:

$$-COOH + -NH_2 \longrightarrow -CONH- + H_2O$$

Acid   Amine   Amide

$$NH_2CH_2CH_2CH_2CH_2CH_2CH_2NH_2 + HOOCCH_2CH_2CH_2CH_2COOH$$

1,6-Diaminohexane   Adipic acid

$$\downarrow$$

$$-\!(NHCH_2CH_2CH_2CH_2CH_2CH_2NHCOCH_2CH_2CH_2CH_2CO)\!- + H_2O$$

Nylon 66

Other condensation polymerisations of commercial importance involve the following reactions:

$$-OH + -COOH \longrightarrow -COO- + H_2O$$

Hydroxyl   Acid   (Poly)ester

$$-OH + -NCO \longrightarrow -NHCOO-$$

Isocyanate   (Poly)urethane

There is a useful analogy for condensation polymerisation and that is to let the monomers be represented by boys and girls. These are sent into the playground and told to run around and instructed that they should join hands with a member of the opposite sex. Boy may not join with boy nor girl with girl, so in consequence a strictly alternating boy–girl–boy–girl chain will develop. The analogy illustrates the effect of having a slight excess of one sex, for example boys, where polymerisation will cease due to all the chains being boy-ended and there being no further girls. However, chains with different end units can join together. The method of lowering chain length by adding a compound with one functional group is analogous to there being a few one-handed boys (or girls) who will stop the further growth of the chain end to which they add.

So far we have limited our discussion of polymerisation to compounds which only give linear polymers. If a monomer contains more than one double bond or ring, or more than two functional groups in the case of condensation polymerisation then branching or crosslinking may occur. Perhaps rubber is the best known example of a crosslinked polymer. Natural rubber consists of linear poly *cis*-1,4-isoprene and by using a suitable initiator this compound can be synthesised from isoprene monomer

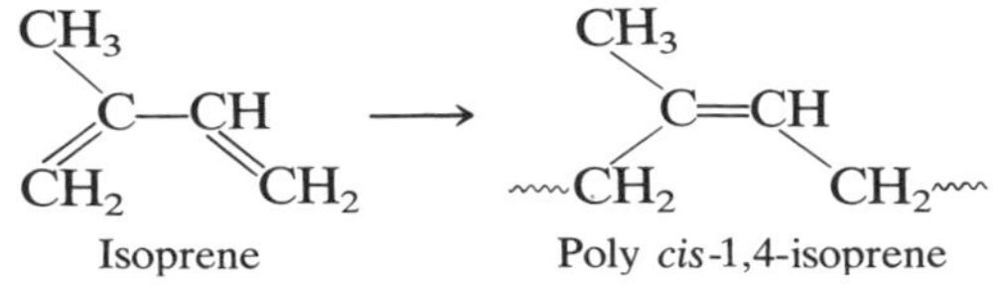

Isoprene   Poly *cis*-1,4-isoprene

A double bond remains in the polymer, and during sulphur vulcanisation this is used to connect neighbouring chains together through short sequences of sulphur atoms. Another well-known crosslinking reaction is the setting of a polyester resin, this is the type of material that might form the matrix in glass-fibre reinforced polyester. Here an unsaturated polyester (short chains containing both ester groups and carbon–carbon double bonds) is dissolved in styrene and a free radical initiator is added. An addition polymerisation takes place involving double bonds in both the styrene and the unsaturated polyester (Fig. 2.3).

The diglycidyl ether of bisphenol A is the basis of the most widely used epoxide resins; it contains two epoxide rings

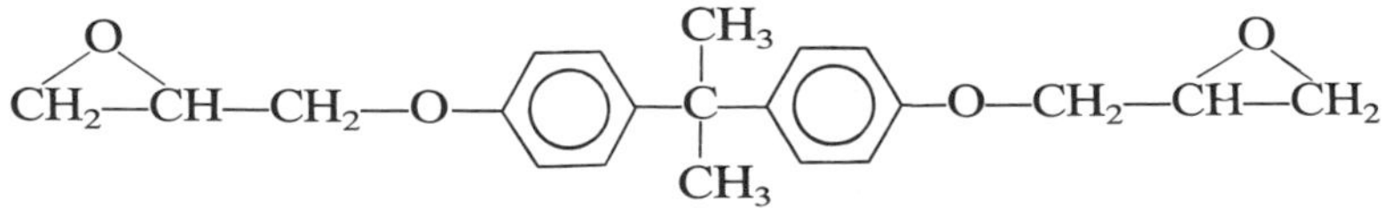

J. Comyn

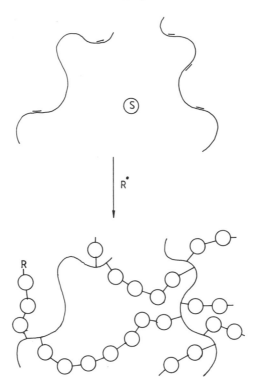

Fig. 2.3. Reaction of free radicals (R˙) with styrene (S) and an unsaturated
polymer polyester to give a crosslinked polymer.

It will react with appropriate hardeners (aromatic amines, aliphatic
amines, acid anhydrides) to produce a crosslinked product. Two hardeners of practical importance are triethylenetetramine and 4,4′-
diaminodiphenylmethane

$NH_2CH_2CH_2NHCH_2CH_2NHCH_2CH_2NH_2$
      Triethylenetetramine

NH_2⟨◯⟩—CH_2—⟨◯⟩NH_2
4,4′-Diaminodiphenylmethane

These react with the diglycidylether of bisphenol A in a condensation
polymerisation on the basis of one amine hydrogen atom reacting with
one epoxide group.

$$\text{>N—H} + \overset{\displaystyle O}{\overset{\diagup\diagdown}{CH_2—CH—}} \longrightarrow \text{>N—CH}_2\overset{\displaystyle OH}{\underset{|}{—CH—}}$$

This means that triethylenetetramine has six reactive groups and 4,4'-diaminodiphenylmethane has four.

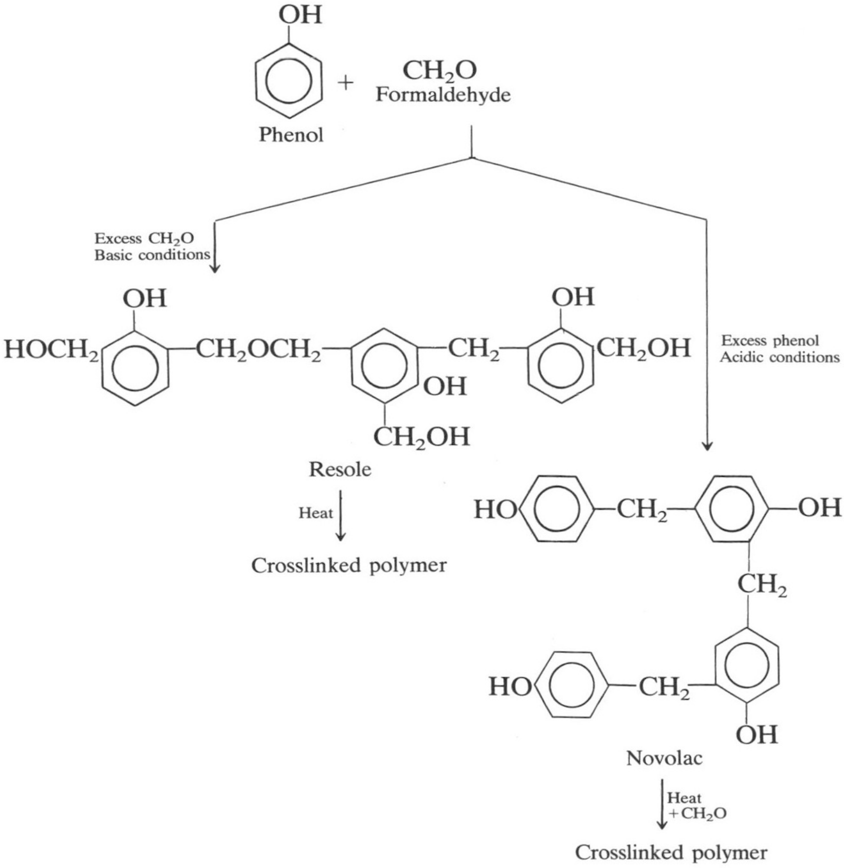

Phenol–formaldehyde resins have been widely used for brown and black electrical fittings, and they represent a thermosetting condensation polymer. Depending upon the ratio of phenol to formaldehyde and the pH of the reaction mixture a resole or a novolac is formed by the substitution of formaldehyde molecules at different positions around the benzene ring. In the case of resoles, phenol groups are bridged by $-CH_2-$ and $-CH_2OCH_2-$ units and methylol groups $-CH_2OH$ occur on the rings. During the moulding process further bridging occurs via the methylol groups and a crosslinked polymer results. The significant difference with novolacs is that methylol groups are absent, and further formaldehyde in the form of hexamethylene-tetramine is added before moulding.

More detailed treatments of addition and condensation polymerisations can be found in Refs 3–6.

2.2. MOLAR MASSES OF POLYMERS

The molar mass of a polymer is its mass when compared with the oxygen 16 atom. Its degree of polymerisation is the number of monomer units it contains. If the mass of the monomer unit is M_0 then the polymer molar mass M is related to degree of polymerisation x by

$$M = M_0 x \qquad (2.1)$$

During the synthesis of linear addition polymers, molar masses are controlled by the free radical lifetime and the rate constant for propagation. Since there is a distribution of radical lifetimes the resulting polymer will have a distribution of molar masses. Methods of measuring molar masses of polymers are not discussed here, but molar masses themselves are important in that most polymer properties depend upon them.

If a polymer contains N_1 molecules of mass M_1, N_2 molecules of mass M_2, etc., with the ith component having N_i molecules of mass M_i, then the number average molar mass $\bar{M}_n$ is given by

$$\bar{M}_n = \frac{\sum N_i M_i}{\sum N_i} \qquad (2.2)$$

Similarly, the weight average molar mass is

$$\bar{M}_w = \frac{\sum N_1 M_1^2}{\sum N_1 M_1} \qquad (2.3)$$

The relationship of these two averages to a typical molar mass distribution curve is shown in Fig. 2.4, where it can be seen that the ratio $\bar{M}_w/\bar{M}_n$ is a measure of the spread of molar masses. It is thus important to remember that any quoted polymer molar mass is an average.

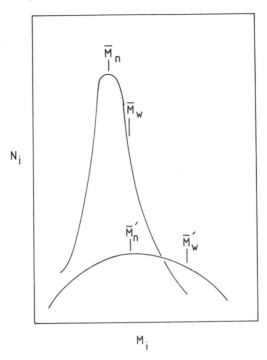

Fig. 2.4. Molar mass distribution curves for polymers of relatively narrow and broad distribution.

The property which shows the strongest dependence on molar mass is melt viscosity, which is obviously an important parameter in the melt processing of plastics. Melt viscosity is a measure of the ease with which chains can slip over each other, and as a chain becomes longer the attractive forces between it and its neighbours increase. There is also an increased possibility of chain entanglement with increasing length.

Further information on molar masses of polymers can be found in Refs 3–5.

2.3. THE PHYSICAL STRUCTURE OF POLYMERS

The analogy between polymer chains and spaghetti is a useful one in examining the physical structure of polymers. As everyone knows,

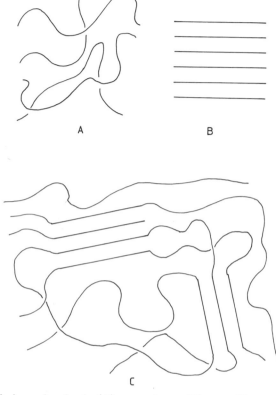

Fig. 2.5. Chain molecules in (A) amorphous, (B) crystalline and (C) semicrystalline regions of polymers (fringed micelle model).

spaghetti can be purchased in a tin where the strands are disorganised,
or in a packet where the straight strands lie parallel in an organised
manner. The two forms of spaghetti are simple models for an amorphous polymer and a crystalline polymer. However, things are more
complicated than this since whilst polymers can be totally amorphous,
they cannot be totally crystalline. Many polymers are hence semicrystalline and a useful though incorrect model for semicrystalline polymers
is the fringed micelle model (Fig. 2.5). Here the amorphous phase
is continuous and individual crystallites are embedded in it. Individual
polymer molecules can have parts of their length in the amorphous
phase and others in crystalline regions.

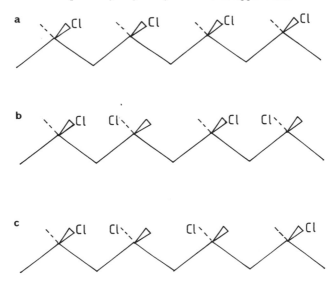

Fig. 2.6. Illustration of the structures of (a) isotactic, (b) syndiotactic and (c) atactic polymers using PVC as an example. Here the carbon–carbon backbone lies in the plane of the page, the dotted bond behind and the triangular bond in front of the page. Hydrogen atoms are omitted for clarity.

A regular chemical structure permits a polymer molecule to crystallise and an irregular chemical structure makes crystallisation impossible. With vinyl and other polymers based on double-bond compounds, this irregularity can take the form of atacticity. If polyvinyl chloride is used as an example to illustrate the subject of tacticity, then three structures are possible for the polymer chain and these can be described by looking at the position of the chlorine atom (Fig. 2.6). If the carbon–carbon backbone is arranged as a fully extended zigzag then the possibilities are an isotactic molecule where all chlorine atoms are on one side, a syndiotactic molecule where they rigidly alternate from side to side, and an irregular or atactic molecule. Free radical initiators tend to give atactic and therefore totally amorphous polymers and for this reason the commercially available forms of polystyrene, polymethylmethacrylate, polyvinyl acetate and others are without crystallinity. Using some ionic or Ziegler–Natta initiators[3,5,6] it is possible to synthesise isotactic or syndiotactic polymers, and catalysts of the latter type are used to make polypropylene and poly-4-methyl pentene-1 both of which are isotactic and consequently semicrystalline in their normal commercial form.

Polyethylene lacks a substituent group and hence tacticity has no meaning here. However, this absence means that polyethylene has a regular structure which may crystallise.

Isotactic and syndiotactic polymers are termed stereoregular, and syndiotactic molecules enter the crystal in the way already described, that is in the extended zigzag form with the chains lying parallel. The reason why this is not the case with the isotactic molecule is that the substituent group is too big to be accommodated in the extended zigzag structure and so the molecule forms a helix, where the bulky substituent groups now radiate from the helix.

The linear condensation polymers nylon 66, nylon 610 and polyethylene terephthalate have regular structures with symmetry about the chain axis and so may crystallise. However, the latter polymer is an interesting case in that slow cooling from the melt gives a semicrystalline polymer, but rapid cooling produces an amorphous material. This illustrates the importance of cooling rate on degree of crystallinity. The common crosslinked polymers have irregular structures and so are amorphous.

The importance of crystallinity in the present context is that it affects the properties of materials. Whilst the amorphous regions of polymers are permeable to oxygen, water, carbon dioxide etc., the tighter packing of crystalline regions makes them impermeable, so that in a semicrystalline polymer the crystallites are impermeable barriers in the path of diffusing molecules.

Crystalline materials such as metals and ice have a sharp melting point above which they are fluid. The same is essentially true of semicrystalline polymers, and for the purposes of melt processing they are heated to temperatures somewhat above the crystalline melting point, T_m. On the other hand the T_m represents an upper service temperature for thermoplastics. Some values of T_m are given in Table 2.1.

Whilst T_m is a property of the crystalline phase in polymers, the glass transition temperature T_g is a property of the amorphous phase. On passing above T_g a polymer changes from being relatively hard and brittle to being soft and leathery, and these changes are associated with the onset of motion of the polymer backbone at T_g. On returning to the tinned spaghetti model, the best analogy for the glass transition is for the sauce to freeze as we pass down through the T_g. However, a more realistic model would be a box full of snakes who suffer from stiff backs at low temperatures, but who began to move about once temperatures exceeded T_g. Some values of glass transition temperatures also appear in Table 2.1.

Different polymers find application in both the glassy and leathery states but the glass transition temperature is usually a thing to be avoided in service. Examples of glassy polymers in service include polymethylmethacrylate in the coloured light lenses in automobiles, and the polyester matrix resin in glass-fibre reinforced plastics, whilst examples of the leathery state in service includes all rubbers, and plasticised polyvinyl chloride in flexible boots and clothing. The undesirability of a change of state in these examples seems obvious.

A semicrystalline polymer will thus exhibit both a glass and a melting transition on being heated up from low temperatures and changes in volume and thermal properties are associated with these. If we were to measure the volume or heat content of a sample of such a polymer against temperature, data similar to those shown in Fig. 2.7

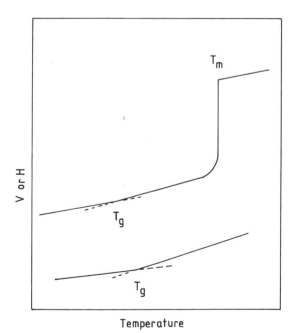

Temperature

Fig. 2.7. Changes in the volume or heat content of a semicrystalline (top) and an amorphous polymer (bottom) with temperature.

would result. At the melting point a sharp change in volume or heat occurs, whereas at T_g the sharp change is not in V or H but in the first derivative dV/dT or dH/dT, so leading to the classification of these changes as first-order (T_m) and second-order (T_g). Measurement of T_m and T_g is often by following changes in volume or heat content, frequently using the instrumental techniques of thermomechanical analysis, differential scanning calorimetry or differential thermal analysis. Intermolecular and intramolecular forces have an effect on the value of both T_g and T_m. Both transitions involve the loosening of a structure held together by intermolecular forces, and the flexibility of the molecule in the higher temperature state will be greater if intramolecular forces are low. These forces will be changed if the structure of a polymer is altered, and any change which, say, has an increasing effect on T_g will also tend to increase T_m. A consequence of this is that for many polymers T_g is between one half and two thirds of the value of T_m. Apart from high density polyethylene and polytetrafluoroethylene, the polymers in Table 2.1 conform well to this guideline.

The glass transition temperature of a polymer may be lowered by the addition of a liquid, an effect which is known as plasticisation. The addition may be chosen or adventitious, respective examples of these being phthalate plasticisers for PVC and atmospheric water causing plasticisation of hydrophilic polymers such as the nylons and epoxides. One theoretical approach to the glass transition which can be presented simply is the free volume theory. A polymer consists of occupied volume and free volume, and upon thermal expansion the amount of free volume increases. Once the amount of free volume reaches a critical amount, polymer chain segments begin to move and the polymer passes through T_g. Hence any action which decreases free volume would have the effect of increasing T_g. Conversely increasing free volume lowers T_g.

As fluids have a large free volume they have a plasticising effect upon polymers. Based on the assumption that the free volume of the mixture is the sum of the free volumes of the components

$$\frac{1}{T_g(PF)} = \frac{w_P}{T_g(P)} + \frac{w_F}{T_g(F)} \qquad (2.4)$$

eqn (2.4) relates the glass transition temperature of the polymer–fluid mixture $T_g(PF)$ to the weight fractions of the two components w_P and w_F and their glass transition temperatures $T_g(P)$ and $T_g(F)$. This equation was first proposed by Fox[7] for binary copolymers and has been used successfully for some polymer–fluid mixtures, but the glass transition temperature of a fluid is not such a straightforward parameter as that of a homopolymer. This is because most common fluids freeze to crystalline solids rather than forming amorphous glasses. Atmospheric water may plasticise polymers. In spite of the ability of water to form ice crystals at 0°C it has a glass transition temperature in the region −134 to −138°C which can be measured by a number of techniques.[8] The effect of water on the glass transition temperature of some epoxide resins[8] is shown in Table 2.2. All T_g values are lowered on first absorbing liquid water and agreement of the experimental values with those calculated from eqn (2.4) is probably within experimental error for the hardeners DAPEE, DAB, DMP and boron trifluoride monoethylamine. In other cases there is less plasticisation than predicted and this may mean that some water is isolated in droplets within the polymer. An item of further interest from this table is that on immersion in water for 10 months, the T_g of every resin rises, and in two cases this is to a temperature above the T_g of the original dry polymer. Here it seems that further crosslinking reactions may occur on prolonged exposure to water.

Table 2.2

Glass Transition Temperatures of Some Epoxide Resins Formed from the Diglycidyl Ether of Bisphenol A with Various Hardeners[8]

Hardener	Measured T_g (°C)			T_g (°C) from eqn (2.4)
	Dry	Wet	After 10 months in water	Wet
Di-(1-aminopropyl-3-ethoxy)ether (DAPEE)	67	37	49	44
Triethylenetetramine (TETA)	99	86	111	76
1,3-Diaminobenzene (DAB)	161	143	157	139
4,4′-Diaminodiphenyl-methane (DDM)	119	110	130	92
Tris(dimethylamino-methyl)phenol (DMP)	68	51	54	47
$BF_3C_2H_5NH_2$	173	155		151

We can return to the theme of the free volume theory to examine the effect of other factors on T_g. The introduction of polar groups into a polymer leads to an increase in intermolecular forces and hence a decrease in free volume; here an example is the high T_g of PVC (81°C) compared with polypropylene (−8°C). On the other hand, the inclusion of non-polar side groups of increasing size causes a progressive lowering of T_g. They have the effect of holding the chains apart and reducing the efficiency of packing the molecules and hence bringing an increase in free volume. The polyalkylmethacrylates

$$-CH_2-\overset{\overset{\displaystyle CH_3}{|}}{\underset{\underset{\displaystyle COOR}{|}}{C}}-$$

illustrate this effect ($T_g = 105$°C when R = −CH$_3$, 65°C when R = −CH$_2$CH$_3$, 35°C when R = −CH$_2$CH$_2$CH$_3$ and 20°C when R = −CH$_2$CH$_2$CH$_2$CH$_3$). Lowering the molar mass of a polymer leads to a lowering of T_g. Chain ends disrupt chain packing and more free volume is associated with them than with chain units. As end units are more plentiful in short chains a larger amount of free volume will occur. Another way of introducing end units is to branch the polymer and hence the effect of branching is a lowering of T_g. Crosslinking has the effect of pulling two chains together and so decreasing the free volume and increasing T_g.

A fuller treatment of the free volume theory has been given by Meares.[9] More detailed treatments on polymers in the solid state appear in Refs 3–5.

2.4. THE EFFECT OF OXYGEN AND ULTRA-VIOLET LIGHT ON POLYMERS

The natural environment interacts with polymers in a manner which is undesirable. Few raw polymers are capable of withstanding the prolonged effects of exposure to UV or oxygen, and the step that is taken to improve their resistance to these agents is to compound them with suitable additives such as antioxidants or UV stabilisers.

Just as polymerisation is a free radical process, the oxidative degradation of polymers has a free radical mechanism, so antioxidant stabilisers can consist of molecules which react with and deactivate free radicals. Hindered phenols such as di-*t*-butyl-*p*-cresol react with radicals in the following manner

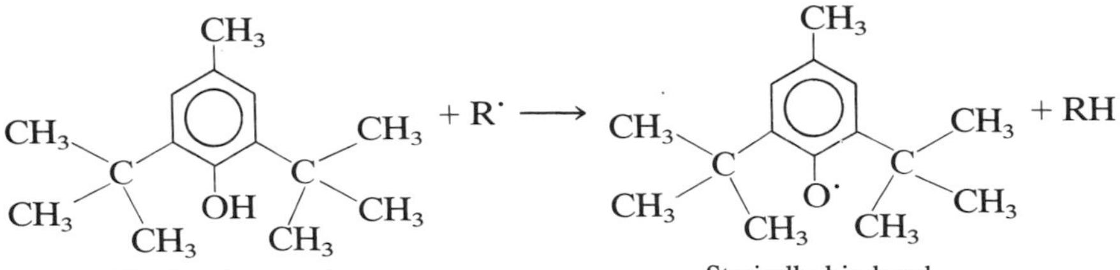

Di-*t*-butyl-*p*-cresol

Sterically hindered
free radical

However, the new free radical cannot react because it is hindered from doing so by the very bulky *t*-butyl groups. Hindered phenols of this type are frequently used as antioxidants for polyethylene and polypropylene to which they are added in quantities of about 1%.

The quantum energy of UV photons is sufficient to break many chemical bonds and the general effect of UV upon polymers is to generate free radicals which then behave as the active centres for polymer degradation. Hence compounds capable of removing or deactivating free radicals will play a role in preventing both oxidative and UV degradation. However, with UV there is also the possibility of adding UV absorbers as stabilisers. Carbon black is an efficient UV absorber and its addition as a filler to vehicle tyres no doubt contributes much to their environmental stability, but has the obvious disadvantage of limiting the choice of colour to black and giving opacity.

UV stabilisers should absorb light without the formation of free radicals, and should absorb wavelengths below 420 nm, as the maximum sensitivity for many plastics is between 290 and 360 nm. Useful absorbers include 2-hydroxybenzophenones which absorb UV photons thus in the hydrogen bonds

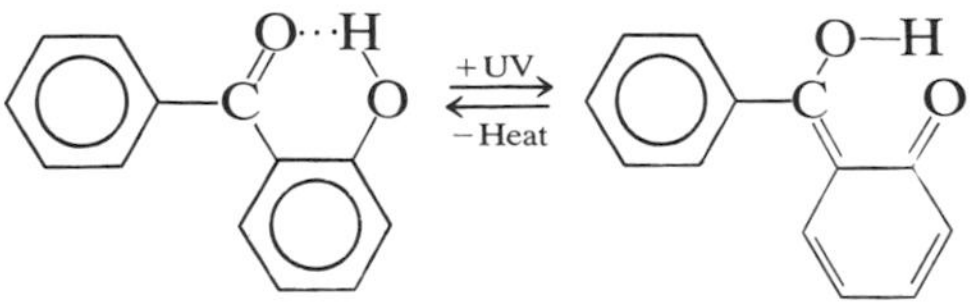

2-Hydroxybenzophenone

Additive free polyvinyl chloride is a particularly unstable polymer which thermally degrades with the liberation of the very corrosive gas hydrogen chloride

$$-CH_2-\underset{\underset{Cl}{|}}{CH}-CH_2-\underset{\underset{Cl}{|}}{CH}- \longrightarrow -CH=CH-CH_2-\underset{\underset{Cl}{|}}{CH}- + HCl\uparrow$$

The double bond remaining in the polymer has the effect of activating the neighbouring C–Cl bond so that it is more likely to react next. The effect of this process, apart from the production of HCl is the formation of neighbouring double bonds in PVC

$$-CH=CH-CH_2-\underset{\underset{Cl}{|}}{CH} \longrightarrow -CH=CH-CH=CH- + HCl\uparrow$$

Such bonds are termed conjugated and they are a source of colour in organic molecules. Thus as degradation proceeds PVC discolours through yellow, orange and brown and finally to black. One group of stabilisers which is used with PVC is the organotin compounds such as dibutyl tin dilaurate, which probably works by exchanging activated chlorine atoms for a butyl group

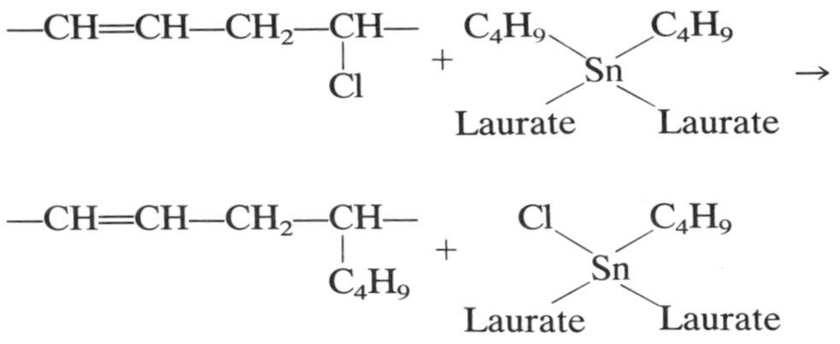

$$-CH=CH-CH_2-\underset{\underset{Cl}{|}}{CH}- \quad + \quad \underset{Laurate \quad Laurate}{\overset{C_4H_9 \quad C_4H_9}{\diagdown Sn \diagup}} \quad \longrightarrow$$

$$-CH=CH-CH_2-\underset{\underset{C_4H_9}{|}}{CH}- \quad + \quad \underset{Laurate \quad Laurate}{\overset{Cl \quad C_4H_9}{\diagdown Sn \diagup}}$$

The stabilisation of polymers is an extensive and complex subject, and here only a few selected samples have been looked at in a fairly simple way. Stabilisers are not the only compounds which are added to polymers, and it is important to remember that polymers are rarely pure but may contain additives for the following reasons:

(1) Processing aids—e.g. lubricants.
(2) Stabilisers—e.g. antioxidants, antiozonants, UV stabilisers, fungicides.
(3) Colourants—dyes or pigments.
(4) Reinforcing agents—particulate fillers, e.g. carbon black, fibrous fillers, e.g. glass or carbon fibres.
(5) Surface property modifiers—antistatic agents, antiwear additives, lubricants.
(6) Plasticisers—e.g. phthalate esters for PVC.

A fuller treatment of the role of additives in polymers is given by Mascia.[10]

2.5. ELECTRICAL PROPERTIES OF POLYMERS

The effect of the electric field upon a polymer could be to cause ionic or electronic conductance, dielectric loss or breakdown. Common polymers are good insulators, some values of volume resistivity appear in Table 2.3, but problems can arise with hydrophilic polymers such as the polyamides. Nylon 66 has a specific resistance of about 10^{15} Ω cm when dry, but on equilibration with saturated air at room temperature this substance absorbs about 8% of water and its specific resistance falls to about 10^9 Ω cm. When a polymer is placed in an electric field, the effect is to displace the centres of gravity of electronic and nuclear charges so that the material becomes dielectrically polarised. Further

Table 2.3

Electrical Properties at Room Temperature and at Low Frequency of Some Polymers

Polymer	Permittivity	Tan δ	Dielectric strength $(MV\,cm^{-1})$	Volume resistivity $(\Omega\,cm)$
Polyethylene	2·3	10^{-4}–10^{-3}	5·3	$>10^{16}$
Polypropylene	2·3	3×10^{-4}–10^{-3}	0·24	10^{16}–10^{17}
Polymethyl methacrylate	3·6	0·62	0·14	$>10^{15}$
Polyvinyl chloride				
Unplasticised	3·5	0·031	0·24	10^{15}
Plasticised	6·9	0·082	0·27	10^{13}
Polystyrene	2·5	$1·5 \times 10^{-4}$	0·2–0·3	10^{17}–10^{19}
Polyethylene terephthalate	3·3	$2·5 \times 10^{-3}$	2·95	10^{18}
Nylon 6 (dry)	3·5	$6·5 \times 10^{-3}$	1·5	10^{15}
Nylon 66 (dry)	3·6	$8·5 \times 10^{-3}$	1·5	$>10^{15}$
Polytetra-fluoroethylene	2·1	2×10^{-4}	0·18	$>10^{15}$
Phenol–formaldehyde resin				
General purpose	6·0–10·0	0·1–0·4	0·06–0·12	10^{10}–10^{12}
Low electrical loss	4·0–6·0	0·03–0·05	0·10–0·14	10^{11}–10^{14}
Typical epoxide	4·5–5·5	0·01–0·02	0·2	10^{14}–10^{15}
Silicone rubber	3·6	2×10^{-3}	0·2	10^{16}

polarisation occurs if there are polar groups present as these will tend to be orientated by the field. Atomic polarisation arises from the movement of atoms or groups of atoms and Maxwell–Wagner polarisation associated with the build up of charge carriers at an impermeable interface such as a filler particle or a crystallite. The relaxation times associated with these processes are around 10^{-13} s for atomic polarisation and at progressively longer times for dipolar and Maxwell–Wagner processes, respectively. Hence a polymer could show a number of loss processes on the application of a very wide range of electrical frequencies, as exemplified in Fig. 2.8. Clearly in any application it is important to avoid the loss of power and generation of heat which follows from operating near a loss peak.

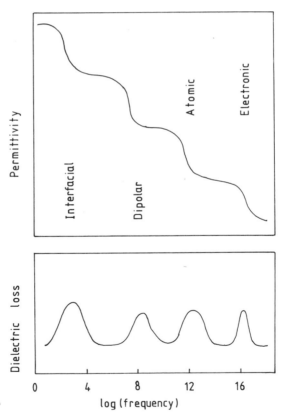

Fig. 2.8. Permittivity and dielectric losses in a polymer over a wide range of frequencies.

Before the advent of PVC for wire insulation, rubber was used. Because rubber contains double bonds (Section 2.1) it is a fairly reactive material, and reaction with atmospheric oxygen is reputed to have introduced a dipolar loss peak at about 50 Hz, which was held responsible for many house fires.

As micro-circuits become smaller, the field strength between conductors increases and may approach levels at which insulators will simply break down. Dielectric strengths of some polymers are included in Table 2.3. Blythe[11] has written a monograph on the electrical properties of polymers. Although the book by McCrum *et al.*[12] first appeared in 1967, it still represents a valuable review of the dielectric properties of polymers.

2.6. BARRIER PROPERTIES

All polymers are permeable to atmospheric gases and vapours, and so cannot be used to make hermetic seals. If a film of polymer separates two chambers containing a gaseous permeant at partial pressures p_1 and p_2 (Fig. 2.9) where $p_1 > p_2$ then the gas will permeate through the film from chamber 1 to chamber 2. In order to do this it will dissolve in the polymer at one face and diffuse through the film to the other face

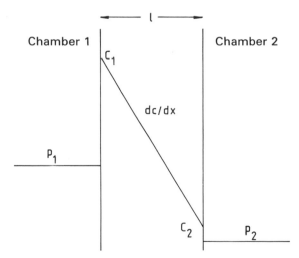

Fig. 2.9. Permeation of a gas through a polymer film in the steady state.

where it will evaporate. Two assumptions need to be made: firstly that the concentration of gas dissolved in the facial layers of the film is proportional to partial pressures at those faces (eqns (2.5) and (2.6))

$$C_1 = Kp_1 \qquad (2.5)$$

$$C_2 = Kp_2 \qquad (2.6)$$

where K is the solubility or partition coefficient; secondly that diffusion through the film is controlled by Fick's 1st law of diffusion (eqn (2.7))

$$F_x = -D \, dc/dx \qquad (2.7)$$

where F_x is the amount of substance diffusing across a unit plane perpendicular to the x-axis in unit time, D is the diffusion coefficient and dc/dx is the concentration gradient. Hence the amount of substance diffusing across a film of area A in time t is

$$Q = -DAt \, dc/dx \qquad (2.8)$$

Clearly

$$dc/dx = (C_2 - C_1)/l = (p_2 - p_1)K/l \qquad (2.9)$$

whence

$$Q = -DAKt(p_2 - p_1)/l \qquad (2.10)$$

Experimental arrangements often mean that $p_1 \gg p_2$ whence

$$Q = -DKAtp_1/l \qquad (2.11)$$

The product of diffusion and solubility coefficients DK is identified as the permeability coefficient P.

Fick's 1st law can only be applied to the steady state, that is where there is neither a build up nor a decay of diffusant in the polymer. A steady state holds in Fig. 2.9 as the concentration–distance plot in the film remains constant. By considering the sum of the fluxes into and out of a volume element (Fig. 2.10), it can be shown that the build up of diffusant in the volume element is given by Fick's 2nd law (eqn (2.12)).

$$\frac{\partial c}{\partial t} = D\left(\frac{\partial^2 c}{\partial x^2} + \frac{\partial^2 c}{\partial y^2} + \frac{\partial^2 c}{\partial z^2}\right) \qquad (2.12)$$

Again experimental conditions often simplify matters by restricting diffusion to one direction (the x-direction) whence the equation sim-

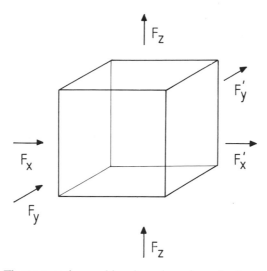

Fig. 2.10. Fluxes entering and leaving a box-shaped volume element.

plifies to

$$\frac{\partial c}{\partial t} = D \frac{\partial^2 c}{\partial x^2} \tag{2.13}$$

Solutions to this differential equation for a number of boundary conditions are available in Crank's text *The Mathematics of Diffusion*[13] and in Carslaw and Jaeger's *Conduction of Heat in Solids*.[14]

If a film of polymer of thickness 2ℓ is immersed in an infinite bath (one which is not altered by absorption by the film) of diffusant then concentrations C at points within the film are given by eqn (2.14). Here x is the space coordinate; the faces of the film are located at $+\ell$ and $-\ell$ and $x = 0$ is the centre of the film

$$\frac{C}{C_1} = 1 - \frac{4}{\pi} \sum_{n=0}^{\infty} \frac{(-1)^n}{2n+1} \exp\left[\frac{-D(2n+1)^2 \pi^2 t}{4\ell^2}\right] \cos\left[\frac{(2n+1)\pi x}{2\ell}\right] \tag{2.14}$$

C_1 is the equilibrium concentration of diffusant and t is time. Equation (2.14) also applies to the situation where a barrier coating of thickness ℓ coats an inert substrate. Here the substrate–coating interface is located at $x = 0$ and the face of the film is at $x = +\ell$. Some values of C/C_1 calculated from eqn (2.14) are shown in Fig. 2.11. The fact that

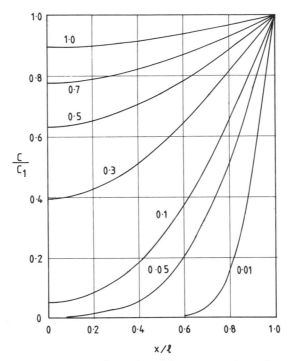

Fig. 2.11. Concentration profiles of a diffusant entering a layer of polymer, thickness ℓ, coating an inert substrate, at several values of Dt/ℓ^2; calculated from eqn (2.14).

all the loci meet at $x/\ell = 1$ is due to the assumption that equilibrium is instantaneously attained at the surface layer. Concentrations of diffusant at the protected interface are given by the values at $x/\ell = 0$ where the equation is simplified by the cosine term equalling unity.

Equation (2.14) can thus be used to calculate concentration of permeants at the protected interface provided the diffusion coefficient and equilibrium solubility are known. It can be handled adequately by personal computers, and users will find that under most conditions the exponential term rapidly diminishes with n, so that only a few values of n need to be considered.

Integration of eqn (2.14) yields a function (eqn (2.15)) which gives the total amount of diffusant taken up

$$\frac{M_t}{M_\infty} = 1 - \sum_{n=0}^{\infty} \frac{8}{(2n+1)^2 \pi^2} \exp\left(\frac{-D(2n+1)^2 \pi^2 t}{4\ell^2}\right) \qquad (2.15)$$

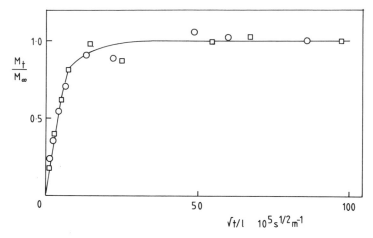

Fig. 2.12. Uptake of water at 45°C by an epoxide resin formed from the diglycidylether of bisphenol-A and tris(dimethylaminomethyl)phenol. Results of two experiments are shown.[15]

The value of this equation is that it provides a simple way of measuring diffusion coefficients for cases where there is sufficient mass uptake for weighing. Here M_t is mass uptake at time t and M_∞ is the equilibrium mass uptake. If mass uptake is plotted against the square root of time the resulting curve typically shows an initially linear region (Fig. 2.12), and from the slope of this region the diffusion coefficient can be obtained using eqn (2.16) which is a short-time approximation from eqn (2.15)

$$\frac{M_t}{M_\infty} = \frac{4}{\ell} \left(\frac{Dt}{\pi} \right)^{1/2} \tag{2.16}$$

Uptake plots of the type shown in Fig. 2.12 are typical of diffusion of substances in leathery polymers (i.e. $T > T_g$) and are termed Fickian. This is in contrast to non-Fickian diffusion which is generally expected for diffusion of organic solvents in glassy polymers. Plot C in Fig. 2.13 shows a typical non-Fickian plot. Here the solvent begins to plasticise the polymer and lower its T_g, so that once T_g is lowered to the experimental temperature there is a sharp increase in the rate of uptake.

Polycarbonate is a polymer which is particularly sensitive to the effect of solvents and some features of its behaviour are shown in Fig. 2.13. In plot A non-Fickian uptake is followed by a region of weight

J. Comyn

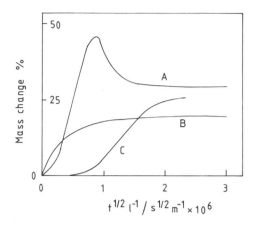

Fig. 2.13. Non-Fickian diffusion processes illustrated by the interaction of polycarbonate with methyl methacrylate vapour at 25°C. Sequential experiments are (A) sorption, (B) desorption in vacuum and (C) resorption.[16]

loss which is due to solvent-induced crystallisation. The loss is due to the expulsion of solvent from the crystalline region. Plot C shows non-Fickian uptake by the new semicrystalline polymer, and comparison of these two plots illustrates that diffusion is slower in the semicrystalline polymer.

Some values of transport parameters for gases in polymers are collected in Table 2.4, and values for water in polymers in Table 2.5.

The atmospheric gases O_2, N_2 and CO_2 have low solubilities in polymers, and the amount of dissolved gas is insufficient to alter the properties of the medium. Fickian diffusion would be expected in such circumstances. Although epoxides are glassy polymers and can absorb significant amounts of water, diffusion of water in them seems invariably Fickian.[15] Certainly it seems reasonable to expect that diffusion of atmospheric components in common insulating and encapsulating polymers is Fickian.

All the equations used above are for Fickian diffusion, non-Fickian diffusion is much more complex.

The reason why polymers are permeable is their relatively high level of molecular motion. Motion of the chain segments can allow a diffusing molecule to pass, and if we return to the spaghetti model then diffusion is the passage of ball bearings through spaghetti. Here the spaghetti strands will be displaced by the moving balls but will reform

Table 2.4

Permeability and Diffusion Coefficients of Atmospheric Gases
and Vapours in Some Polymers at 25°C

Polymer	Gas	$P \times 10^{10}$ $(cm^3\ STP)\ cm$	$D \times 10^{-10}\ (m^2\ s^{-1})$
		$\overline{cm^2\ s\ (cm\ Hg)}$	
High density			
polyethylene	O_2	2·88	0·46
	CO_2	12·6	0·37
	N_2	0·969	0·32
	H_2	90	
Low density			
polyethylene	O_2	0·403	0·170
	CO_2	0·36	0·24
	N_2	0·143	0·093
	H_2O	12·0	
Polytetra-			
fluoroethylene	O_2	4·2	0·152
	CO_2	11·7	0·095
Polyvinyl			
chloride	O_2	0·0453	0·0118
	CO_2	0·157	0·00256
	H_2O	275	0·0238
Nylon 66	H_2O	0·0069	0·00083
Polyethylene			
terephthalate			
Crystalline	O_2	0·035	0·0035
	N_2	0·0065	0·0014
	CO_2	0·17	0·0006
Amorphous	O_2	0·059	0·005
	N_2	0·013	0·002
	CO_2	0·30	0·00085

behind them. In Section 3 it was mentioned that the glass transition
marks the onset of chain segmental motion and large increases in
diffusion coefficient might therefore be expected once T_g has been
exceeded.

Manson and Chiu[19] measured the water permeability of some films
of an epoxide resin over a range of temperatures and found that plots
of log (permeability) against reciprocal temperature consisted of two

Table 2.5

Diffusion   Coefficients   and   Solubilities   of   Water   in   Some
Polymers[8,17,18]

Epoxides with the following hardeners[a]	$t(°C)$	$D(m^2 s^{-1})$	Equilibrium uptake (%)
DAPEE	25	$1\cdot3\times10^{-13}$	5·0
TETA	25	$1\cdot6\times10^{-13}$	3·8
DAB	25	$1\cdot9\times10^{-13}$	2·3
DDM	25	$9\cdot9\times10^{-15}$	4·1
DMP	25	$2\cdot0\times10^{-13}$	4·4
$BF_3C_2H_5NH_2$	25	$1\cdot6\times10^{-13}$	2·3
FM 1000 (epoxide-polyamide adhesive)	25	$1\cdot1\times10^{-12}$	15·0
	50	$3\cdot2\times10^{-12}$	21·0
Epoxide novolac encapsulant	50	$1\cdot2\times10^{-10}$	1·2
Silicone rubber	50	$4\cdot4\times10^{-11}$	0·2
PMMA	50	$8\cdot2\times10^{-12}$	0·66

[a] The epoxide resin used was based on the diglycidyl ether of bisphenol-A.

straight-line regions which intersected at about T_g. The activation energy for permeation was greater for the leathery polymer. A similar relationship for the activation energies of diffusion was reported by Meares[20] for the diffusion of helium, neon, argon, oxygen and hydrogen in polyvinyl acetate; he accounted for the lower activation energy in the glassy state by there being a smaller zone of activation. The zone of activation can be thought of as the region of a polymer which needs to be disturbed to accept a diffusing molecule (making a hole in the spaghetti into which a ball bearing can move).

Due to tight packing the crystalline regions of polymers are impermeable. Any diffusing molecule is thus excluded from the crystallites and these present barriers around which a diffusant must pass; increasing crystallinity thus lowers the apparent diffusion coefficient and is thus a way of improving barrier properties. The diffusant only dissolves in the amorphous phase so the apparent solubility is also diminished. The effect of a filler particle is much the same as crystallites, assuming of course that preferential transport does not occur at the filler–polymer interface by capillary action. Apart from their high level of molecular motion making polymers permeable in comparison to other substances such as metals, glasses and ceramics, it also makes them much more susceptible to creep. This again uses the spaghetti model which sees strands sliding over one another during creep. Knotting the strands together (i.e., crosslinking) is clearly a way of reducing creep.

Whilst there is a moderate body of information on the transport of small neutral molecules in polymers, the same cannot be said of ion transport in polymers. Clearly ion mobility is a factor of the greatest importance in the encapsulation and insulation of electrical components. Most of our knowledge on this topic probably comes from the science of dyeing.[21] Cellulose, nylon and wool absorb anionic dyes (reactive dyes, direct dyes, vat dyes in leuco form) from aqueous solutions. These first three types of dye contain $-SO_3^-$ groups and are thus strong electrolytes (i.e., highly dissociated) in water. However, what is probably the most important issue is that these fibres absorb large quantities of water from the dyebath. The amounts absorbed from saturated air at ambient temperatures are 42% by weight for regenerated cellulose, 32% for wool, 24% for cotton and 8% for nylon 66. Water has the effect of plasticising the fibres and so increasing the diffusion coefficient, but perhaps more importantly it increases the permittivity of the medium (polymers have low permittivities (Table 2.3) whilst water has a permittivity of about 80 at room temperature) and so allows a significant number of dye anions to remain dissociated. Another feature of dyeing is that in some fibres there are sites of opposite charge to the dye ion; these include $-NH_3^+$ sites in acidified wood, $-O^-$ sites in alkaline cellulose and $-SO_3^-$ or $-SO_4^-$ sites in acrylic fibres. These provide sites for the location of dye ions, but may also be essential for dye diffusion by providing 'stepping stones'. Thus it may be the case that for ions to be mobile in a polymer we need either a high permittivity to reduce aggregation or sites of opposite charge to act as 'stepping stones'.

2.7. ADHESION PROPERTIES

Adhesion of an encapsulant to the substrate is a desirable property in that it helps to prevent the transport of moisture by capillary action along the interface. A number of polymers are used as adhesives and these include the following types:

(1) *Reactive adhesives* Here the monomers are applied to the substrate. Their low viscosity allows them to develop intimate

contact with the surface. A strong adhesive bond develops as the adhesive polymerises. Examples of such adhesives are epoxides, cyanoacrylates and reaction-setting acrylics.

(2) *Solvent-based adhesives* The adhesive is a polymer solution, and again its low viscosity assists wetting of the surface. As the solvent evaporates, a tacky adhesive layer is left behind. Emulsion-based adhesives are similar to these but here droplets of adhesive stabilised by soap molecules are dispersed in water. A considerable disadvantage of solvent-based adhesives is the toxicity and flammability of many organic solvents, but to replace them by water is not a simple matter as the high heat of evaporation leads to slow solvent removal.

(3) *Hot melt adhesives* These are linear thermoplastics which are applied to the substrate in the melt. Joint strength develops on cooling. They have low toxicity and are amenable to automation. Examples are copolymers of ethylene and vinyl acetate, some polyesters and some polyamides.

The efficiency of an adhesive may be assessed by making joints and measuring the load needed to break them using a suitable mechanical testing instrument. However, this is not a simple matter because firstly there are a large number of experimental variables to select and control, and secondly the stress distribution in joints is fairly complex. Experimental variables may include the following: (i) treatment of surfaces to be bonded; (ii) control of preparation and application of the adhesive; (iii) joint geometry including control of glue line thickness; (iv) pressure and temperature of bonding; and (v) control of rate of stress or strain during testing. There are a large number of joints used in laboratory testing and some of these are shown in Fig. 2.14. One problem is that there are three principal stresses in such joints, these are as illustrated in Fig. 2.15. The single lap joint contains all types of stress; obviously there is an in-line shear stress, cleavage stress arises due to cylindrical bending of the adherends, and a Poissons ratio shrinkage on stretching the adherends leads to the sideways cleavage (Fig. 2.16). Further, these stresses are not evenly distributed within the joint; they are concentrated along the overlap edges of the joint. Thus although the testing of adhesive joints is very useful in comparing materials or processes, it is very difficult to translate between joints of different types.

A further complication in the use of adhesives is that whilst rigid

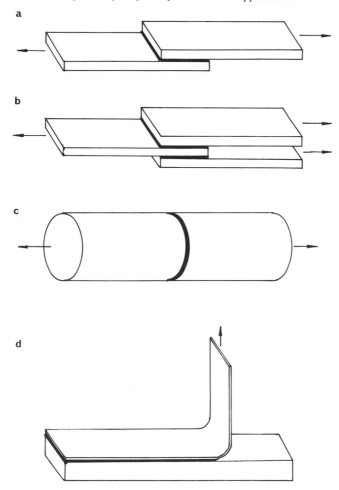

Fig. 2.14. Some joints used in laboratory testing of adhesives. (a) Single lap, (b) double lap, (c) butt, and (d) 90° peel joints.

adhesives (e.g., epoxides) are resistant to shear stresses, their resistance to cleavage is poor. The opposite is true for rubbery adhesives. Thus in an attempt to produce adhesives which are strong in both peel and shear, rubber-modified rigid adhesives have been developed, where the cured adhesive is a crosslinked block copolymer.

There are a number of current theories of adhesion and these have

J. Comyn

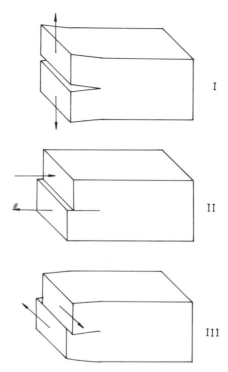

I

II

III

Fig. 2.15. The principal modes of fracture. Mode I in cleavage or peel, Mode II in-line shear and Mode III sideways shear.

been reviewed by Kinloch[22] and by Wake.[23,24] The mechanical inter-locking theory proposes that the keying of the adhesive into surface irregularities of the substrate is the major source of adhesive strength. Although the surfaces of solids are never perfectly flat, the surface irregularities of a substance like glass are small; however, where this method of attachment is probably important is in the bonding of polymers to textiles, and in the adhesive bonding of wood, where adhesive penetrates into the pore structure. The diffusion theory considers that polymers adhere to each other by diffusion of chain segments across the interface. It therefore presupposes that the chain segments possess sufficient mobility and that they are mutually soluble. Whilst the first of these conditions would be met by polymers in the leathery state different polymers are not mutually soluble. Even polymers which are chemically very similar such as polyethylene and

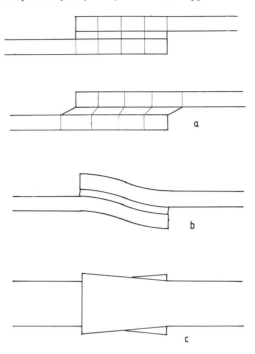

Fig. 2.16. Deformation of a single lap joint on tensile loading. (a) Elements in the lap area are stretched and the glue line deforms giving maximum distortion and stress towards the edges of the lap. (b) Bending of the lap area occurs giving a maximum cleavage stress at the edges of the lap. (c) The adherends shrink laterally on stretching to give a Mode III stress which again is greatest at the edges of the lap.

polypropylene are mutually insoluble so the theory probably only applies to bonding a rubbery polymer to itself. Such a case would exist in the solvent welding of heat-sealing plastics. In the former, solvents are used to weld glassy thermoplastics such as polymethyl methacrylate or polycarbonate where they have the effect of introducing molecular motion by their plasticising action.

If the adhesive and substrate have different electronic band structures there is likely to be a transfer of electrons between them and a resulting double layer of electrical charge at the interface. Derjaguin and Smilga[33] have proposed that the electrostatic forces across the interface are a major source of joint strength, but this is questioned by Wake.[24] In a number of cases there may be specific interactions at the interface. An example is an adhesive containing copolymerised acid bonded to a metallic (oxide) substrate where an acid–base interaction might occur. It has also been suggested that ion-pairs may be formed at the interface between epoxide adhesives and aluminium pretreated in acid baths.

The adsorption theory invokes the use of van der Waals' forces to provide sufficient bond strength. Although these are the weakest of intermolecular forces (Table 2.6) their anticipated combined strength is much greater than can be obtained in adhesive joints. Two types of force are involved here, dispersion forces occur between all molecules and are due to the attraction between non-permanent dipoles. If permanent dipoles occur then the stronger polar forces occur. Plastic substances such as polyethylene, polypropylene and polytetrafluoro-ethylene are of very low polarity and their bondability can be greatly improved by introducing polar chemical groups by surface treatment.

The idea of a weak boundary layer is a theory of non-adhesion rather than one of adhesion. However, weak boundary layers certainly occur on some substrates. Examples include lubricating oil on metals and rust on iron, but a less obvious case lies in the tendency of foreign material in plastics to diffuse to the surface; such materials might be the various additives discussed in Section 2.4 or simply low molar mass polymer. In either case the remedy is to remove the weak boundary layer with a solvent or by abrasion.

Certainly surface treatment is a valuable way of producing good adhesion. The subject has recently been reviewed in a book edited by

Table 2.6
Typical Energies of Bonds and Intermolecular Forces[25]

Bonds	Energy $(kJ\,mol^{-1})$
Interionic	580–1000
Covalent	60–700
Metallic	110–350
Hydrogen bonds involving fluorine	up to 40
Other hydrogen bonds	10–25
van der Waals' forces	
Dipole–dipole	4–20
Dipole-induced dipole	<2
Dispersion forces	0·08–40

Brewis.[26] Abrasion and solvent wiping have already been mentioned, but other techniques involve treating plastics in a gas flame, corona discharge for the treatment of both plastics and metals, and the use of acid-etch baths (chromic or phosphoric acid) for metals and plastics or anodising in acid baths for metals.

Aluminium has been more widely studied as a substrate for adhesive bonding than any other metal, because of its use in building aircraft. Here the major problem is to produce joints which are durable to atmospheric water vapour. Some data on the strengths of aluminium single lap joints on exposure to wet air are shown in Fig. 2.17. It can be seen that the strengths of dry joints from the various surface treatments are virtually the same, but that significant changes develop on exposure. Anodising pretreatments in chromic–sulphuric or phosphoric acid give the best durability with aluminium and their role is to produce a stable oxide layer. The role of water in the weakening of adhesive joints has been reviewed.[15]

Surface treatments in strong acid solutions seem highly undesirable for aluminium conductors which are narrow and thin. However, vacuum-deposited aluminium should have the virtue of being clean, although it will be covered with a not particularly durable spontaneous oxide. A surface treatment which might be of value here is the use of silane coupling agents.

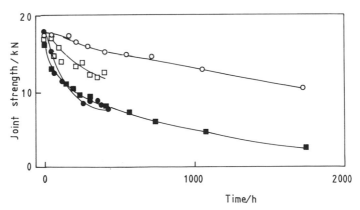

Fig. 2.17. Strengths of some single lap joints of aluminium bonded with an epoxy–polyamide adhesive on exposure to air at 43°C and 90% relative humidity for various times. Metal surface treatments are etching in: chromic–sulphuric acid, ○; alkaline etching, □; solvent degrease, ■; and phosphoric acid etching, ● (after Butt and Cotter[27]).

Silane coupling agents were first developed for pretreating glass fibres prior to their incorporation in organic matrices, and a range of such compounds is now commercially available. When applied to a substrate they are hydrolysed by atmospheric water and consequently polymerise and adhere to a glass surface. The fact that glass itself contains silicon–oxygen chains is seen as a reason for the success of these primers on glass; the established chemical principle of similar materials having a natural affinity is being obeyed. The functional group is then available for chemical reaction with the matrix (amine or epoxide group for epoxide matrices and vinyl group with unsaturated polyesters) so that a covalent bond is formed between primer and matrix. Whatever the mechanism for silane primers this technology certainly works, for without it glass-fibre reinforced plastics would not survive in the hulls of the various vessels, from small dinghies to minesweepers, which employ it.

There has recently been renewed interest in silanes together with experimental evidence that they are effective on surfaces other than glass. Gettings and Kinloch[28] showed that the silane priming of steel can improve the resistance of adhesive joints to water and that this improvement is directly related to the detection of the $Fe-O-Si^+$ ion by secondary ion mass spectrometry. Walker[29-31] has demonstrated their value in pretreating a number of metals (aluminium, mild steel, cadmium, copper and zinc) prior to the application of paints. Silane coupling agents are covered further in Chapter 10. Additional information on adhesion can be found in Refs 22, 23 and 24. Details on surface treatment and analysis appear in Ref. 26 and on silane coupling agents in a recent book by Plueddemann.[32]

REFERENCES

1. Brydson, J. A., *Plastics Materials*, 4th edn, Butterworth Scientific, London, 1982.
2. Brandrup, J. and Immergut, E. H., *Polymer Handbook*, Wiley-Interscience, New York, 1975.
3. Elias, H. G., *Macromolecules*, 2 volumes, Plenum Press, New York, 1977.
4. Young, R. J., *Introduction to Polymers*, Chapman and Hall, London and New York, 1981.
5. Cowie, J. M. G., *Polymers: Chemistry and Physics of Modern Materials*, Intertext Books, Aylesbury, 1973.
6. Odian, G., *Principles of Polymerization*, 2nd edn, John Wiley and Sons, London, 1981.

7. Fox, T. G., *Bull. Amer. Phys. Soc.*, **1** (1956) 123.
8. Brewis, D. M., Comyn, J., Shalash, R. J. A. and Tegg, J. L., *Polymer*, **21** (1980) 357.
9. Meares, P., *Polymers: Structure and Bulk Properties*, Van Nostrand Co. Ltd., London, 1965.
10. Mascia, L., *The Role of Additives in Plastics*, Edward Arnold, London, 1974.
11. Blythe, A. R., *Electrical Properties of Polymers*, Cambridge University Press, Cambridge, 1979.
12. McCrum, N. G., Read, B. E. and Williams, G., *Anelastic and Dielectric Effects in Polymeric Solids*, John Wiley & Sons, London, 1967.
13. Crank, J., *The Mathematics of Diffusion*, 2nd edn, Oxford University Press, Oxford, 1975.
14. Carslaw, H. S. and Jaeger, J. C., *Conduction of Heat in Solids*, 2nd edn, Oxford University Press, Oxford, 1959.
15. Comyn, J., In: *Durability of Structural Adhesives*, (A. J. Kinloch (Ed.)), Applied Science Publishers Ltd, London, 1983.
16. Cope, B. C., PhD thesis, Leicester Polytechnic, 1977.
17. Brewis, D. M., Comyn, J., Cope, B. C. and Moloney, A. C., *Polymer*, **21** (1980) 345.
18. Goosey, M. T., PhD thesis, Leicester Polytechnic, 1982.
19. Manson, J. A. and Chiu, E. H., *J. Polym. Sci.*, *Symposium No. 41*, p. 95 (1973).
20. Meares, P., *J. Amer. Chem. Soc.*, **76** (1954) 3415.
21. Peters, R. H., *Textile Chemistry. Vol. III, The Physical Chemistry of Dyeing*, Elsevier, Amsterdam, 1975.
22. Kinloch, A. J., *J. Mater. Sci.*, **15** (1980) 2141.
23. Wake, W. C., *Polymer*, **19** (1978) 291.
24. Wake, W. C., *Adhesion and the Formation of Adhesives*, 2nd edn, Applied Science Publishers Ltd, London, 1982.
25. Good, R. J., In: *Treatise on Adhesion and Adhesives. Vol. 1 Theory*, (R. L. Patrick (Ed.)), Edward Arnold, London, 1967.
26. Brewis, D. M., *Surface Analysis and Pretreatment of Plastics and Metals*, Applied Science Publishers Ltd, London, 1982.
27. Butt, R. I. and Cotter, J. C., *J. Adhesion*, **8** (1976) 11.
28. Gettings, M. and Kinloch, A. J., *J. Mater. Sci.*, **12** (1977) 2511.
29. Walker, P., *J. Coat. Tech.*, **52** (1980) 49.
30. Walker, P., *J. Oil Col. Chem. Assoc.*, **65** (1982) 415.
31. Walker, P., *J. Oil Col. Chem. Assoc.*, **65** (1982) 436.
32. Plueddemann, E. P., *Silane Coupling Agents*, Plenum Press, New York, 1982.
33. Derjaguin, B. Y. and Smilga, V. P., *Adhesion, Fundamentals and Practice*, McLaren & Son, London, 1969.

Chapter 3

SILICONE PROTECTIVE ENCAPSULANTS AND COATINGS FOR ELECTRONIC COMPONENTS AND CIRCUITS

J. H. DAVIS

Dow Corning Ltd, Barry, Wales, UK

3.1. DEFINITION AND HISTORY

Silicones are synthetic polymers based on a molecular structure of alternating silicon and oxygen atoms with organic groups also attached to all or some of the silicon atoms. Their general formula is $[R_a SiO_{(4-a)/2}]_b$ where $a = 1$ to 3 and $b \geq 2$. R represents an organic group.

Diagrammatically, their smallest unit structure can be represented by

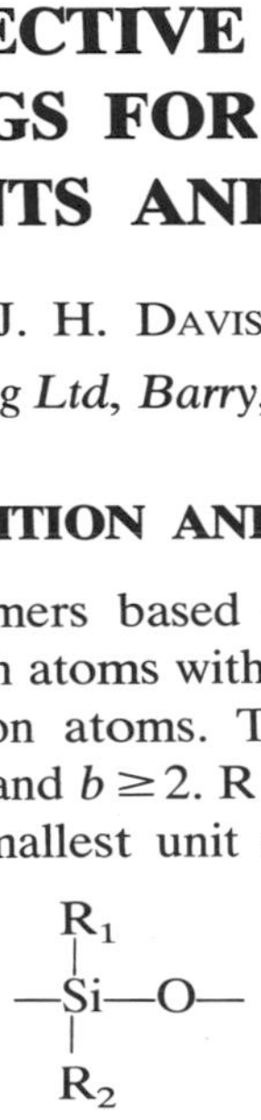

$$\begin{array}{c} R_1 \\ | \\ -Si-O- \\ | \\ R_2 \end{array}$$

where R_1 and R_2 may be equal or dissimilar organic groups. Most commonly these are methyl $-(CH_3)$ or phenyl $-(C_6H_5)$, but may also be chemically reactive groups such as $-CH{=}CH_2$ (vinyl) or $-OH$ (hydroxy) which take part in further polymerisation.

This unit structure makes up the final silicone polymer by molecular chain lengthening and crosslinking, for example

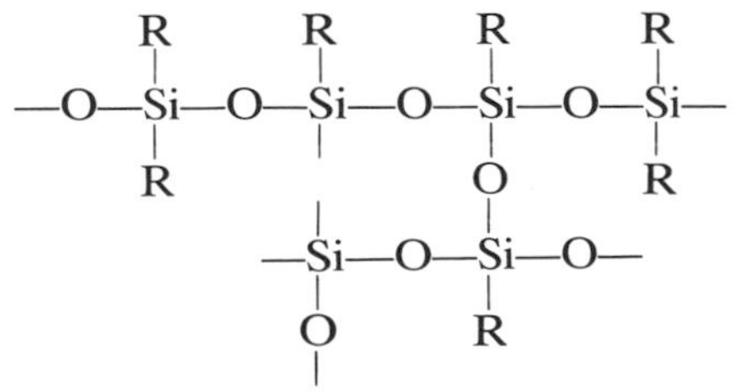

$$\begin{array}{c} R \quad\quad R \quad\quad R \quad\quad R \\ | \quad\quad\ | \quad\quad\ | \quad\quad\ | \\ -O-Si-O-Si-O-Si-O-Si- \\ | \quad\quad\ | \quad\quad\ | \quad\quad\ | \\ R \quad\quad O \quad\quad R \\ \quad\quad\quad | \\ -Si-O-Si-O- \\ | \quad\quad\ | \\ O \quad\quad R \\ | \end{array}$$

67

Physically, silicones are now available as fluids, greases, gels, elastomers of many kinds and hard resins.

The term 'silicone' was first derived by Professor Kipping and his
co-researchers of Nottingham University, England, at the beginning of
this century under the mistaken belief that new silicon-based materials
which they had synthesised were chemically analogous to ketones
based on carbon.[1]

Although the term 'organopolysiloxane' in which the word siloxane
defines the –Si–O–Si– constituent is now considered to be a better
chemical description, the name 'silicone' has lived on and is now
universally used to describe organopolysiloxanes which are used for
industrial applications.

Although Kipping is recognised as the founder of silicone chemistry,
a period of 30 years elapsed before the industrial and development
potential of his work was realised. Corning Glass Works in the USA
had by then developed the first glass fibres for which electrical insulation was foreseen as an important end application. Until then only
organic textile fabrics were available, apart from asbestos, to make
electrical insulating tapes, sheet and laminates. The advent of
glasscloth offered the prospect of a considerable increase in thermal
endurance, subject to the availability of new bonding and impregnating
resins which would also withstand high temperatures.

Work to achieve such polymers based on Kipping's original dis-
coveries began under Dr Frank Hyde at Corning Glass Works. As the
associated chemical technology expanded, Dow Chemical joined as a
partner and in 1943 Dow Corning Corporation was established and the
first commercial production of silicones began. General Electric of the
USA proceeded in parallel and work also began separately in the
Soviet Union. Silicones are now manufactured at large-scale and highly
capital-intensive plants in the USA, Europe and Japan. There are also
several plants carrying out formulation and further processing, although not basic manufacture.

Global annual production (in 1980) has been estimated at 340 000
tonnes valued in total at more than US$2 billion.[2] The valuable
physical properties of silicones described below have led to applications in virtually all industries.

3.2. PHYSICAL PROPERTIES

3.2.1. Thermal and Oxidative Stabilities

The most important characteristic of silicones is their resistance to thermal and oxidative deterioration. This can be largely attributed to the high average bond energy of the Si–O chemical bond, 452 kJ/mol.[3] Although the organic side groups impose temperature limitations, their own thermal endurance is enhanced by the support of the siloxane (Si–O–Si) molecular backbone. Excellent resistance to ultra-violet radiation, ozone, electrical arcing and discharges are also corresponding characteristic features.

3.2.2. Stability in Physical and Dielectric Characteristics with Varying Temperature

The Si–O–Si bond angle in a silicone polymer can vary between 100 and 180° and there is free rotation about these bonds. Such a polymer can adopt a randomly coiled shape which responds easily to external forces unless there are sufficient restraining crosslinks. Lightly crosslinked silicones, even those of high molecular weight, are therefore quite elastic.

Silicones are generally non-polar in behaviour despite the strongly polar Si–O bond. This is attributed to the organic side groups which form a protective sheath preventing the $\overset{+}{Si}-\overset{-}{O}$ dipoles from approaching each other closely. This can be seen from the molecular model of Fig. 3.1 which conveys a better idea of the molecular form than a planar illustration. Because of this separation of the $\overset{+}{Si}-\overset{-}{O}$ dipoles the intermolecular forces between adjacent molecules, following an inverse square law, are also weak. Many characteristics of silicones are due to this behaviour. In organic polymers as the temperature rises the strong

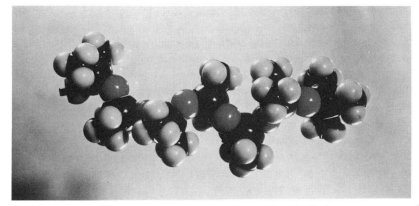

Fig. 3.1. Model of silicone fluid molecule showing helical arrangement of polymer chain with silicon atoms (black) bonded to oxygen atoms (dark grey) and to outer organic groups.

attraction between polymer chains diminishes and dependent properties such as viscosity, elasticity, tensile strength, permittivity and dissipation factor may change significantly. For silicones, intermolecular forces which are already low do not change much as temperature rises or falls, so that physical and dielectric characteristics of the polymer are also relatively stable. Two characteristics of this kind are the relatively high compressibility of silicone fluids and only a small change in their viscosity with temperature variation when compared with mineral oils.[4]

Low and stable values of dissipation factor and permittivity under varying temperatures and also varying frequency became apparent when electronic applications for silicones were first developed over 30 years ago. Von Hippel[5] measured very stable values of dielectric characteristics for silicone resin and silicone resin bonded glasscloth laminates at temperatures up to 200°C and at frequencies from 100 Hz to 100 MHz

A less favourable consequence of these weak intermolecular forces is that silicone elastomers, for example, are physically weaker than many organic rubbers. However, modern polymer and filler technology enables silicone rubbers to be formulated with fully acceptable mechanical strength for industrial uses which are very different from the early materials of 30 years ago.

3.2.3. Surface Properties
The organic side groups in a silicone molecular structure account mainly for characteristic surface properties. Surface tension at 20 mN/m (for methyl silicone fluids) is low and characteristic of a completely organic polymer surface. Such methyl silicones are highly water-repellent. Their low surface tension and insolubility in water make them valuable as antifoam agents and highly effective in breaking down aqueous foams. They also make excellent release agents though by contrast, with reactive side groups present, silicone adhesives can also be formulated.

3.2.4. Handling Characteristics
Silicones are among the least toxic of all polymers. In a few cases they are supplied or applied in common aromatic solvents such as xylene which require normal and appropriate handling precautions. In certain other cases common curing catalysts such as peroxides or organometallic salts are used and these additives again call for the usual safe

handling practice normally prescribed for such materials. In other respects, both when used as production materials by industry and afterwards in service, silicones are extremely safe.

3.2.5. Fire Resistance and Products of Combustion

Silicones are generally very fire resistant, depending on their formulation. Certain silicone elastomers used as encapsulants and cable insulation have excellent values of limiting oxygen index (LOI) or UL rating. A specially processed dimethyl silicone fluid, Dow Corning® 561, used as a transformer coolant is far more fire resistant than mineral oil.

When silicones do burn, a residual ash of silica remains. The heat of combustion is less than for most other polymers, while smoke is also low in density and relatively non-toxic, consisting mainly of carbon dioxide and some carbon monoxide with no corrosive or highly poisonous constituents. These attributes are increasingly important nowadays with attention to more stringent fire regulations.

3.2.6. Chemical Purity and Low Water Absorption

Silicones contain very low levels of impurities such as mobile ions and absorb very little water. As already noted, their surface is highly water-repellent. On the other hand, their open molecular structure makes them more permeable to water vapour than many other polymers. Taken together, these characteristics mean that a silicone-coated substrate will be well protected provided that it is not itself absorbent and that surface contact at the interface is good.

Under these conditions, water vapour may pass relatively freely through the silicone layer in either direction but is not absorbed and does not form a moisture film at the interface. Harmful ionic constituents are not transported to the surface so the protection provided is often superior to that of other polymers which, although less permeable to water vapour, are also more impure chemically.

3.3. PRODUCTION OF SILICONES

The most important method of production in use today begins with a chemical reaction between methyl chloride and metallic silicon at about 300°C, catalysed by copper. This produces a mixture of methyl chlorosilanes which have the general formula $Si(CH_3)_x Cl_{4-x}$. Chlorosilanes have differing boiling points. Unlike the final silicone product, they are highly flammable as well as volatile and must therefore be separated by fractional distillation in an inert atmosphere. A silicone manufacturing plant is characterised by its tall distillation columns.

Methyl chlorosilanes comprise by far the largest volume of silicone precursors but phenylchlorosilanes are also important. One way in which they can be produced is by reaction of chlorobenzene with silicon. Other production processes involve the production of methyl or phenyl chlorosilanes with reactive groups such as vinyl or hydrogen attached to some silicon atoms.

Thus a series of chlorosilane monomers is first produced and then separated by distillation as precursors for conversion to useful polymer products. The next stage is achieved by hydrolysis, in which the chlorosilanes react with water to form compounds called silanols. Hydrogen chloride is also formed and removed:

$$\equiv\text{SiCl} + \text{H}_2\text{O} \longrightarrow \equiv\text{SiOH} + \text{HCl}$$
$$\underset{\text{Chlorosilanes}}{\phantom{\equiv\text{SiCl}}} \qquad\qquad \underset{\text{Silanols}}{\phantom{\equiv\text{SiOH}}}$$

Further reaction of silanols leads to silicone polymers

$$\equiv\text{SiOH} + \equiv\text{SiCl} \longrightarrow \equiv\text{Si}-\text{O}-\text{Si}\equiv + \text{HCl}$$
$$\underset{\text{Silicone}}{\phantom{\equiv\text{Si}-\text{O}-\text{Si}\equiv}}$$

or

$$\equiv\text{SiOH} + \equiv\text{SiOH} \longrightarrow \equiv\text{Si}-\text{O}-\text{Si}\equiv + \text{H}_2\text{O}$$

These equations describe in simplified form a complex production technology that involves a series of processes to purify, stabilise and polymerise the final silicone products which are ultimately available as silicone fluids, gums and resins.

Further processing in which appropriate fillers, catalysts, pigments and other additives are incorporated leads to silicone greases, elastomers and resin-based transfer moulding compounds. A broad list of final silicone products is given in Table 3.1.

Most silicone fluids and silicone greases are used in the form in which they are supplied and require no further polymerisation or curing by the customer. The other products listed, elastomer- and resin-based materials, usually require further curing in the course of extrusion, moulding, encapsulating or similar processes. They are therefore formulated with appropriate reactive chemical groups, as already described, which can be activated by heat, atmospheric moisture or other means to produce the desired end product.

Table 3.1
Typical Silicone Products and Applications

Products	Applications
Fluids	Transformer dielectric. Release agent. Additives for polishes and personal care products. Photocopier fluids.
Reactive fluids	Water repellents for textiles. Release coatings for paper.
Greases	Release. Electrical insulation.
Elastomers	
low consistency	Electronic coatings and encapsulants. Building construction sealants. Mould making
high consistency	Extrusions and mouldings. Cables, seals and gaskets. Telephone and micro-computer keypads. Surgical implants.
Resins	
in solvent	High temperature paints, electrical insulation. Conformal coatings.
solid, powder	Electrical insulation. Transfer moulding compounds for electronic components. Speciality mechanical mouldings

Table 3.2 lists some of the most common of curing reactions. In the first of these (a) the reaction of hydroxyl (OH) groups extends and crosslinks the Si–O–Si chain and is characteristic of many silicone resins, including the resin constituent in silicone transfer moulding compounds which are described later. Usually this reaction is induced by heating the uncured resin which contains some Si–OH groups, in

Table 3.2
Silicone Polymerisation Reactions

(a) $\equiv$SiOH	$+$ HOSi$\equiv$	$\rightarrow \equiv$Si—O—Si$\equiv$	$+$ H_2O
(b) ($\equiv$SiOR)$_2$	$+$ H_2O	$\rightarrow \equiv$Si—O—Si$\equiv$	$+$ 2ROH
(c) ($\equiv$SiOOCCH$_3$)$_2$	$+$ H_2O	$\rightarrow \equiv$Si—O—Si$\equiv$	$+$ 2CH$_3$COOH
(d) $\equiv$SiOR	$+$ HOSi$\equiv$	$\rightarrow \equiv$Si—O—Si$\equiv$	$+$ ROH
(e) $\equiv$SiCH$=$CH$_2$	$+$ HSi$\equiv$	$\rightarrow \equiv$Si—CH$_2$—CH$_2$—Si$\equiv$	

the presence of an organo-metallic salt as a catalyst that is already mixed with the resin as supplied. A small amount of moisture is liberated as a by-product. In the second reaction, (b), alkoxy (RO) groups react with atmospheric moisture liberating alcohol as a by-product. Chemically neutral low consistency elastomers which cure at room temperature (RTV materials) generally fall into this category, as do silicone conformal coatings. To protect against moisture during storage such materials are supplied in tubes, cartridges or other sealed containers. Again there is usually an integral catalyst to promote these curing reactions. Unlike most other curing mechanisms this type of reaction depends on the relative humidity of the atmosphere and cannot be greatly accelerated by heating. (A faster heat cure is possible with some silicone conformal coatings if a special heat-sensitive catalyst is added.)

Because reaction (b) is chemically neutral there is no corrosion hazard affecting electronic equipment. However, the most common family of one-component RTV elastomers employs the acetoxy reaction shown in (c) which liberates acetic acid with its characteristic vinegary smell when curing. This type of elastomer is widely used for general industrial and household sealants but should be avoided in favour of the neutral variety for most electronic applications because acetic acid residues can promote corrosion and undesired electrical conductivity.

The neutral alkoxy curing reaction of (b) can also be incorporated in a two-component system, leading to a cured polymer by interaction between Si–OR and Si–OH groups as in (d), a type of cure often used with two-component RTV products. Like the related reaction (b), curing cannot be appreciably accelerated by heating but no atmospheric moisture is required. The requisite OH groups are contained in one component of the polymer. This permits curing in depth, whereas the one-component systems requiring atmospheric moisture are only suitable for coatings which are a few millimetres in thickness.

Addition reaction (e) is used with many of the two-component silicone elastomers and gels. Silicon-hydrogen groups react with silicon–vinyl groups in the presence of an integral catalyst such as a platinum complex. Unlike reaction (c), there are no by-products and their absence eliminates or minimises shrinkage during cure. Products can be designed to have either very long or very short pot lives as desired after mixing the two constituent components. Elastomers cured by addition reaction (e) are also extremely resistant to reversion or depolymerisation, which may occur under adverse high temperature conditions in confined thick sections of alkoxy cure system elastomers.[6]

A further curing reaction is commonly used for high-consistency silicone rubbers which are fabricated into insulating components such as extruded insulation sleeving, cables and moulded components. Reaction between vinyl and methyl or methyl with other methyl groups is brought about by a peroxide catalyst under heat and pressure using hot air, a heated mould or steam pressure as appropriate.[6]

3.4. APPLICATIONS

The different types of silicone and their application in the electronics industry are now reviewed with illustrative examples.

3.4.1. Silicone One- and Two-Component Elastomeric Sealants and Encapsulants

A wide range of one-component silicone sealants is available for general industrial use but for many electronic applications the acetoxy one-component system should be avoided as already indicated. The properties of four representative products are listed in Table 3.3. The first of these (1) is a free-flowing paste while the second (2) is more thixotropic. Where high cohesive strength is required, the third material shown (3) is preferred particularly for sticking components on to PCB wiring boards and other substrates. The fourth material (4) is used both as a sealant and as a conformal coating, for which purpose it is diluted if necessary with an aromatic solvent to reduce its viscosity.

All these products cure by reaction with atmospheric moisture. Account must be taken of this in application and so they are used for coatings up to about 6 mm thick but are unsuitable for bonding large impermeable areas of material together unless adhesive contact only at the outer edges is sufficient.

Figure 3.2 illustrates the use of sealant 4 to bond ferrite components to a conventional epoxy–glass circuit board. In this case, the upright ferrite cores are partially embedded and locked in place by the sealant, while a fibreglass interface tape is also bonded by a silicone adhesive to the PCB substrate. This provides additional resilience and cushioning protection for the total assembly permitting operation from -55 to

Table 3.3

Typical One- and Two-Component Silicone Sealants, Encapsulants and Conformal Coatings

Silicone Product	Description	Colour	Viscosity (mPa s)	Specific gravity	Pot life (h)	Cure (h/°C)	Hardness (Shore A)	Tensile strength (MPa)	Elongation (%)	Tear strength (kN/m)	Shrinkage (%) 7 days at 25°C	Water absorption (%)
1	Solventless elastomer	Clear/white	25 000	1·04	—	24/25	35	2·0	300	3·6	—	0·4
2	General purpose	White	Paste	1·04	—	72/25	25	1·1	400	2·8	—	0·4
3	High strength	Clear/grey	Paste	1·12	—	72/25	33	4·7	675	2·2	—	0·4
4	Solventless elastomer	Clear	35 000	1·05	—	24/25	25	2·0	350	3·6	—	0·4
5	Flame retardant	Black	3 000	1·38	1	8/25	55	3·4	150	1·7	0·1	0·1
6	Self-priming	Black	1 500	1·23	>4 days	1/150	40	1·0	70	—	0·1	—
7	General purpose	Clear	5 000	1·05	2	24/25	40	6·2	100	2·7	0·1	0·1
8	Self-adhesive gel	Clear	330	0·97	2	24/25	—	—	—	—	—	—
9	Elastoplastic resin solution in solvent	Clear	800	1.1	—	24/25	40(Shore D)	3·0	40	—	—	0·01

Silicone product	Description	Colour	Thermal conductivity (W/m K)	$CTE \times 10^4$ (1/K) (volume)	Weight loss (%) 96 h at 200°C	Electric strength (kV/mm)	Permittivity	Dissipation factor	Volume resistivity (ohm cm $\times 10^{15}$)
1	Solventless elastomer	Clear/white	0·21	—	6·2	24	2·7	0·002	1
2	General purpose	White	0·21	9·3	8·5	20	2·9	0·001	3
3	High strength	Clear/grey	0·17	7·8	7·1	24	2·8	0·003	0·5
4	Solventless elastomer	Clear	0·12	8·8	7·0	20	2·6	0·001	0·5
5	Flame retardant	Black	0·34	8	2·9	18	3·1	0·002	1
6	Self-priming	Black	—	8·5	—	18	3·0	0·002	0·5
7	General purpose	Clear	0·15	9·6	1·6	24	2·6	0·001	2
8	Self-adhesive gel	Clear	—	—	2	17	2·9	0·001	2·3
9	Elastoplastic resin solution in solvent	Clear	0·12	6·3	—	48	2·7	0·001	4·5

Fig. 3.2. Ferrite memory cores locked in place on silicone rubber coated glass-cloth with an elastomeric silicone adhesive. The coated glass-cloth forms the top layer of a rigid epoxy–glass printed circuit board, acting as a further buffer against relative expansion and mechanical shocks (Ampex Corp.).

+100°C at 90% relative humidity without inducing magneto-restriction in the cores.

A one-component silicone RTV (in this case the acetoxy type which can be used because no corrosion problems arise) can be applied to screwheads in a high-power loud-speaker. This prevents speaker rattling due to loosened screws, completes the requisite damping against speaker vibration or handling shocks and has proved more effective than the use of lock-washers. It was also found that a thin layer of this RTV sealant around the edge of the speaker cone considerably extends its life. Remaining resilient, the sealant strengthens the paper and has no effect on the normal speaker vibration characteristics.

Two-component elastomeric encapsulants can provide protection in depth and offer many opportunities for electronic packaging design where resistance to high or low temperatures and physical resilience are required. In general, the cured materials require further support from mechanical damage by an outer container unless there is an external cover. Thus they can best be described as potting materials. Unlike their one-component counterparts, which are normally applied as pastes or viscous fluids from a tube or cartridge, two-component silicone elastomers are supplied as fluid polymers which are mixed before use. Degassing under vacuum is often necessary after mixing to remove air. The mixed material is then poured into a mould or container holding the circuit to be protected and cured at room temperature or by heating if an accelerated cure is desired in the case

Fig. 3.3. High voltage cascade units for VDU terminals arranged in the vacuum vessel in which the circuits are encapsulated in a two-component silicone elastomer. (Wigo GmbH and IBM Corp.)

of addition reaction products. These procedures are relatively simple and can be carried out in a laboratory or workshop with little equipment while for larger-scale use machinery is available to mechanise production with airless mixing, dispensing and curing on a continuous basis.

Table 3.3 also lists the properties of a representative selection of two-component products. The first of these (5) is grey-black in colour and easily handled because the two liquid components as supplied are mixed in equal weights or volumes. They are white and black, respectively, so that complete mixing is apparent when a uniform dark grey colour is obtained.

Figure 3.3 shows the use of this encapsulant to insulate and protect high-voltage cascade units for visual display terminals, with processing in a fully automated vacuum, mixing and transfer system. The silicone encapsulant is highly resistant to fire, passing the stringent UL 94V-0 standard specified by the computer manufacturer end user, and also protects against corona discharge at high voltage.

A primer is sometimes needed with this encapsulant but a new product (6) has been developed with an integral primer, though a heat cure is necessary for its activation. The focus assembly for a colour TV monitor shown in Fig. 3.4 is protected in this way. The 8 kV working voltage is confined to a thick film circuit on a ceramic base which is mounted in turn on a conventional printed circuit board. The applica-

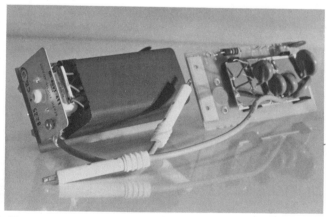

Fig. 3.4. Focus assembly for television monitor in which a hybrid circuit is encapsulated by a self-priming two-component silicone encapsulant. (Tektronix Inc.)

tion of a primer by spray prior to encapsulation proved difficult because of the narrowness of the gaps between hybrid and circuit board but the problem was solved by using this new self-priming product. At the same time, 6 h were saved on the previous processing schedule when a primer had been used before encapsulation.

The next product (7) is a transparent encapsulant, again processed by mixing two fluid components as supplied. One version will cure either at room temperature or by heating, while a variant has much longer pot life at room temperature and is normally cured by heating in the temperature range 70–150°C. The cured product is optically clear, permitting visual inspection of an encapsulated circuit or components. One recent application, outside the definition of electronic encapsulation, is the protective coating of optical fibres.[7,8] In a continuous process the silicone is cured in a few seconds as the fibre passes through a radiant curing zone at over 300°C. The absence of any

Fig. 3.5. Injection of a silicone gel into chip carriers. (AMP Inc.)

solvent or by-products prevents blistering and the silicone provides a chemically pure coating which is also soft enough to prevent mechanical damage to the fibre.

Product (8) is a silicone gel. This has unusual and particularly valuable properties as a protective encapsulant, because in the cured form it is thixotropic and jelly-like in consistency but also has the property of self-healing. Surrounding an encapsulated circuit, the gel continuously maintains surface contact preventing formation of a moisture film at the interface and rehealing if physical contact is broken.

Use of silicone gel in this way is illustrated in Fig. 3.5. Chip carriers to JEDEC design are filled with the uncured liquid gel by meter-mix application in-line and heated in a tunnel through which the conveyor strip passes to cure the gel. This then provides protection to the chip and its leads against both corrosion and mechanical stress caused by differential thermal expansion of the package components.

Silicone gels have also been used effectively to protect automotive electronic modules in the severe environment under a vehicle bonnet,[9] where hazards include high temperature, water and salt from road surfaces.

3.4.2. Conformal Coatings and Impregnating Resins

Unlike the encapsulants described above, conformal coatings provide a relatively thin protective cover of up to about 0·8 mm for printed and hybrid circuits, or other similar configurations. At this thickness, the coating can protect against surface moisture and pollution and thus against unwanted current leakage or tracking, but is not expected to provide the same in-depth mechanical protection or high voltage insulation that is given by a thick section encapsulation.

Within these constraints conformal coatings provide an economical way to ensure circuit reliability. Several different classes of polymer may be used in this way, among which silicones offer the unique combination of a wide temperature working range, surface water resistance and low water absorption.

Two types of silicone are available. Each is formulated to 'conform' as the generic name implies by following the circuit contours without leaving sharp solder points or other obtrusions uncovered. Their general properties are also shown in Table 3.3. The first of these (4) is a solventless, viscous fluid elastomer which can be applied directly to a circuit board by flow coating, but is usually diluted by solvent to reduce its viscosity. Application may then be by dipping, spraying or flow coating as convenient. Curing is by reaction with atmospheric moisture at room temperature. Figure 3.6 shows a circuit for airborne electronic equipment about to be dip-coated in this way. For this process the surface adhesion has been increased by pre-treating the circuit with a primer. In this application the silicone coating has provided particularly good protection against conditions of extreme condensation by comparison with the previously used polyurethane coating.

By contrast, the second conformal coating in Table 3.3 (9) can best be described as an elastoplastic resin: a resin with highly elastic properties. Supplied in solvent, it can again be applied in several ways but forms a harder, more slippery surface after cure than the elastomeric formulation. Maximum elongation at 100% is less than for elastomeric product (4) but is adequate for most applications. Curing is

Fig. 3.6. Dip coating a circuit for airborne electronic equipment in a silicone elastomeric conformal coating. (Gould Inc.)

again by reaction with atmospheric moisture at room temperature, although a fast heat cure is also possible if a suitable catalyst is added. The most common application in electronics is to protect printed circuits, and it has been used successfully in this way for automotive applications.[10] In addition, it has proved effective as an impregnating varnish which cures at room temperature for the insulation and protection of small high-voltage power supply transformers. By keeping the uncured silicone out of contact with moisture until the curing stage is reached, long pot life is attained. The finished surface of these components is smooth, tack-free and unlike the previously used wax impregnations does not pick up dust.

A paper by Day and Weller[11] reports the good fire-resistant characteristics of this type of conformal coating. When applied to an epoxy–glass circuit board, the silicone coating did not improve the assembly to the same degree of fire resistance as the silicone itself, but the lower value for the uncoated board was not impaired. Other types of coating reduced the fire resistance of the uncoated epoxy–glass board.

One useful attribute of these two silicone coatings is their repairability. Solder joints can be re-made, often even without removal of the coating. No carbon is formed and fresh silicone may then be applied to the area which has been repaired. Both types also comply with the specification MIL-I-46058C for PCB coatings, and with draft proposals for a new IEC specification. One limitation is their poor resistance to strong solvents such as fluorinated hydrocarbons which are often used in PCB processing.

3.4.3. Transfer Moulding Compounds

Plastic packaging of semiconductors was first commercially developed in the early 1960s and has grown to be the most common form of protection for these components. Silicones played a significant part in this development and continue to be used for the protection of several types of component including transistors of TO 220 and TO 3 configurations, some power diodes and also passive components such as power resistors which must withstand severe temperature cycling and thermal ageing (Fig. 3.7).

However, silicone transfer moulding compounds have only been used to a limited extent for packaging integrated circuits. This is due to the less effective lead seal of the silicone resins compared with epoxies, a factor which is more important for a multi-lead IC than for a discrete device. This lead-seal deficiency can be overcome by a technique

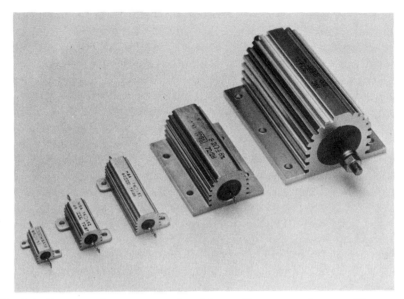

Fig. 3.7. Power resistors encapsulated in a silicone transfer moulding compound. (C.G.S. Resistors.)

known as back-filling[12] in which the moulded devices are vacuum-pressure impregnated in silicone resin to fill the small space between lead and plastic. But this additional process cost is only acceptable for some higher priced professional devices. In addition, the silicone compound, while of adequate strength for a discrete moulding, is measurably lower in flexural and torsional strengths than epoxy in a long, thin DIL package.

Silicones are also more expensive than epoxy compounds but the smooth and fast release of silicone from the mould tool can still save substantially on total production cost.

Unlike applying the elastomeric encapsulants previously described, transfer moulding requires the use of a press capable of pressures up to 3 or 4 MPa and temperatures of about 175°C, in combination with precision multi-cavity steel moulds which may hold several hundred discrete devices. For these reasons, transfer moulding is only justified for high-volume production (see Chapter 5).

A range of silicone transfer moulding compounds is available, but to illustrate their general performance Table 3.4 shows the properties of

Table 3.4
Typical Silicone Moulding Compound Characteristics

	Type reference		
	1	*2*	*3*
Flexural strength (MPa)	60	52	63
Flexural modulus (MPa)	8 200	4 900	10 400
Compressive strength (MPa)	97	69	124
Tensile strength (MPa)	35	31	35
Thermal conductivity (W/m K)	0·63	0·58	0·75
Linear coefficient of thermal expansion $(10^{-6}\,K^{-1})$			
parallel to flow	22	22	25
perpendicular to flow	27	29	26
Electric strength (kV/mm)	12·8	12·4	12·0
Volume resistivity (dry)$(\times 10^{15}\Omega\,cm)$	2·8	2·7	2·8
Volume resistivity (wet)$(\times 10^{15}\Omega\,cm)$	2·8	2·7	2·8
Permittivity (dry)[a]	3·19	3·61	3·60
Permittivity (wet)[b]	3·22	3·64	3·65
Dissipation factor (dry)[a]	0·001 6	0·001 8	0·001 4
Dissipation factor (wet)[b]	0·001 9	0·002 7	0·001 4
Moulding pressure (MPa)	2·7	3·4	3·4
Spiral flow (cm) (5·5 MPa, 177°C)	99	69	69[c]

[a] Measured at 1 MHz.
[b] Measured at 1 MHz after 24 h immersion in distilled water at 23°C.
[c] At 6·9 MPa.

three typical formulations. Product 1 is a general-purpose compound designed primarily for the protection of high power density transistors and similar devices. Product 2 is particularly suitable for high temperature devices, including passive components such as power resistors which with appropriate design can then withstand thermal shock and temperature cycling from −65°C to +350°C. The third product in this table, designated 3, has almost isotropic thermal expansion characteristics and is particularly suitable for smaller signal and power transistors.

Table 3.5 compares the typical ingredients of silicone and organic transfer moulding compounds. Because the silicone is naturally fire retardant, no additive for this purpose is needed, nor does the silicone require as much release agent as an organic compound. Silicone

Table 3.5
Transfer Moulding Compounds—Ingredients

Silicone compound	Organic compound
Mineral fillers	Mineral fillers
Resin	Resin (containing chloride)
Catalyst (non-amine)	Catalyst (sometimes amine)
Pigment	Pigment
Silicone features	Release agent (much larger amount than in silicone compound)
Low chloride	
Low alkali	Fire retardant (often halogenated materials)
Inherently non-flammable	
Inherently moisture resistant	

products listed in Table 3.4 have the following common characteristics:

(1) Low dielectric loss which changes very little with variation in temperature or electrical frequency.
(2) High working temperature capability.
(3) Very low water absorption.
(4) No glass transition in the processing or working temperature range. (Consequently there is no related adverse mechanical effect on a chip or its connections caused by an abrupt change in thermal expansion characteristics during manufacture or in service.)
(5) Fire retardant without additives.
(6) Minimal tool wear, with easy mould release.

3.4.4. Device Performance

Apart from superior performance at high temperatures and under thermal cycling, probably the most valuable feature of silicone packaging is resistance to the effects of moisture at both normal and higher temperatures.[13]

Ionisable impurities in a plastic become mobile in the presence of moisture, particularly at higher temperatures. These impurities tend to move to the chip surface and may cause junction inversion, an increase in leakage current under reverse voltage and also corrosion of aluminium interconnections. To counteract this, silicone plastics have the double merit of low ionic content and low water absorption even though, as already noted, water may permeate more easily along the lead–plastic interface. Epoxy plastics inevitably contain some chloride and have higher water absorptions together with catalyst residues which may be ionic and capable of forming an aggressive electrolyte.

A study of plastic packaging by Lawson[14] showed that a straight-line relationship exists for each type of plastic if the median life to failure for a number of devices is plotted on a logarithmic scale against the square of relative humidity.

More recently, Roberts[15] has derived acceleration factors for heat–damp testing of small silicone packaged signal transistors. He concluded that the majority of silicone packaged devices which he had evaluated under damp heat conditions were superior to those packaged in epoxy with the proviso that other unknown factors relating to the chip or connection design may have partially contributed to this superiority.

3.4.5. High Purity Electronic Coatings and Adhesives

In order to protect a semiconductor surface against adverse physical and chemical effects from its surroundings, and also to increase the breakdown strength at P–N junction surfaces in many devices, silicone polymers of special formulation with optimum rheology and chemical purity have been developed. Most of these are elastomeric, although one type is a hard resin. Typical products are shown in Table 3.6. These high-purity coatings may be applied to a chip surface before encapsulation, for example prior to transfer moulding, or used alone to provide surface protection on a naked chip mounted on a hybrid circuit or in a similar situation.

In these products, the combination of low moisture absorption, very low levels of metallic or chloride ions and compatibility with the chip surface is of key importance. Figure 3.8 illustrates the use of two such coatings in the packaging of axial power diodes and high power TO 220 or TO 3 transistors. The diode chip is protected by a thixotropic silicone paste which fills the remaining space between the adjoining electrodes and thus reduces the inverse voltage leakage current. In the case of the power transistor, a high-purity silicone elastomer forms a protective coating not only for the semiconductor surface but also over the connecting bond wires. It has been found best in practice to cover the bond wire near to the chip surface but not along its whole length, in order to avoid failures due to wire breakage. Similar types of coating are also used to prevent corrosion, current leakage or failure under inverse voltage at thyristor junctions.

The rheology desired for many of these applications includes high

Table 3.6
High-Purity Electronic Coatings

Description	Colour	Viscosity (mPa s)	Specific gravity	Pot life (h)	Cure (h/°C)	Flow (mm)	Solvent	Solids content (%)	Hardness (Shore A)	Electric strength (kV/mm)	Permittivity	Dissipation factor	Volume resistivity (Ω cm $\times 10^{15}$)
Solventless elastomer	White	35 000	1·25	—	2/150	60	—	100	35	16	3·5	0·001	0·1
Solventless elastomer	White	18 000	1·25	—	2/150	90	—	100	35	16	3·5	0·001	0·1
Elastomeric dispersion in solvent	Clear/grey	1 200	1·05	—	72/25	—	Xylene	50	24	23	2·69	0·0015	0·35
Resin solution in solvent	Straw	150	1·05	—	2/150	—	Toluene/xylene	50	—	90	3·12	0·001	30
Solventless elastomer	Clear	6 000	1·09	14 days	2/150	—	—	100	30	22	3·00	0·001	2·0
Solventless elastomer	Black	6 000	1·09	14 days	2/150	—	—	100	30	22	3·00	0·001	2·0

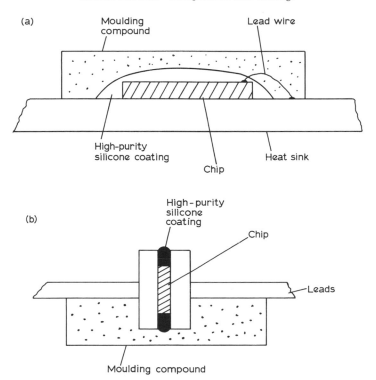

Fig. 3.8. Protection of chip surfaces with high-purity silicone coatings. (a) Transistor; (b) axial diode.

thixotropy, so that shearing forces during dispensing allow the coating to flow closely round the chip surface and bond wires. The original viscosity must then be rapidly restored so that the silicone coating remains *in situ* without further flow during the curing stage, whether at room temperature or by heating at, say, 120–150°C. Figure 3.9(a) shows the dynamic viscosity characteristics of a recently developed silicone blob coating with a marked diminution in viscosity under shear. This particular product is cured by heating, and typical of silicone polymers shows a much smaller fall in viscosity as temperature rises prior to cure than an organic blob coating such as an epoxy (Fig. 3.9(b)). This ensures that the blob does not flow away from the chip during cure.

An example of its application is shown in Fig. 3.10. A circuit

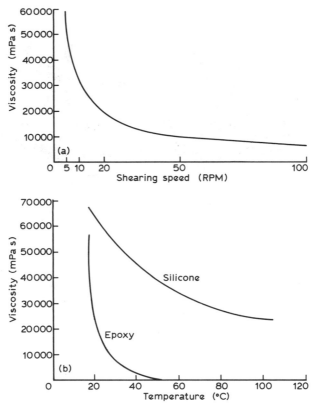

Fig. 3.9. Viscosity characteristics of silicone blob coating. (a) Viscosity characteristic using Brookfield RVT viscometer; (b) comparison between silicone and epoxy.

incorporated in a control unit for small electric motors which are used in hand drills contains a hybrid circuit comprising wire-bonded diodes on a ceramic substrate with an adjacent thyristor. The blob coating does not flow outwards during a 150°C cure, thus avoiding adverse effects on adjoining thick film resistors. Furthermore, the thyristor is coated in a single application without any spillage over the heat sink.[16]

Outside the field of discrete component protection, high-purity silicone coatings contribute to the growing use of VLSI microcircuits combined with newer forms of interconnection and packaging. These include tape automated bonding, beam-lead chips and chip carrier

Fig. 3.10. A silicone blob coating protects semiconductor chips on a hybrid circuit in a control unit for small electric motors. (Capax B.V.)

packaging. Figure 3.11 shows a product line in which a high-purity silicone protective coating is applied by syringe to telecommunication circuits. These range from single chips to circuit packs which contain 50 or more beam-lead integrated circuits, with various combinations of thin film resistors and capacitors and as many as 5000 beam crossovers.[17]

The product is easy to use: as a simple, one-part system it is easily applied by automatic or semi-automatic dispensing equipment. Under vacuum the coating flows to fill the minute voids beneath the beam-lead chips and crossovers. But external leads need not be masked because the coating rheology is such that flow stops at the edge of the ceramic substrate. Curing is by reaction with moist air after evaporation of the xylene solvent. In this case a final cure at 120°C removes residual traces of both solvent and the alcohol by-product prior to final packaging in either a metal case or an epoxy or silicone moulding compound.

The same coating is also used to protect static RAM chips from alpha particle radiation, which can cause soft errors, and it reduces the probable failure rate by at least one order of magnitude.[18] Calculations indicated that a 0·5 mm layer would stop 8 MeV alpha particles, and that the coating was fully compatible in all respects. The authors reported that the resultant soft error rate of 0·1% per 1000 h established in service was the lowest level of activity observed for any packaging material presently in use.

Fig. 3.11. Application of a high-purity silicone coating to protect telecommunication microcircuits. (Western Electric Co.)

Speciality silicone adhesives have also been developed for the varied and critical bonding applications in chip and hybrid circuit mountings. They combine the advantages of chemical purity, temperature cycling capability and elasticity. This enables two rigid surfaces of different thermal expansion coefficient to be stuck together. One newly developed silicone adhesive is available as an easily mixed two-part system which flows easily at room temperature and can be applied in very thin layers to achieve maximum heat transfer to a heat sink from a hybrid substrate or other components.[16]

3.4.6. Insulating and Constructional Materials, Wires and Cables

Many fabricated materials made from or based on silicones are also available for the construction of electronic equipment, including silicone resin bonded glasscloth laminates, insulating sleeving cables or wiring and a wide variety of moulded or extruded components.

Silicone resin bonded glasscloth laminates have low and stable dielectric loss at temperatures up to 200°C or even higher. They are used in the construction of radar and high-power communications equipment, and in dielectric heating apparatus. Copper-clad silicone–glass laminates are available for PCB manufacture, but find application mainly where copper-clad epoxy–glass laminates are unacceptable in terms of thermal endurance or dielectric loss. The copper to laminate peel strength is lower and more variable in the case of silicone than epoxy, and the laminate price is higher.

Silicone elastomer sleeving is available in two forms: unsupported extrusions and with fibreglass reinforcement. The reinforced variety is mechanically stronger but may be lower in electric strength and has lower elongation. Liquid silicone rubbers have been recently developed which cure by an addition reaction as in (e) of Table 3.3. They have proved to be useful as coatings for reinforced sleeving. Compared with the other two-component elastomers in Table 3.3, these liquid silicone rubbers are higher in viscosity before cure and physically tougher afterwards. They can be economically handled by airless mix and meter systems with automated production equipment providing fast throughput.[19]

Electrical insulating tapes made from silicone elastomers or silicone-resin coated glasscloth or mica paper are also widely used. Polyester or polyimide tapes are also available with heat-resistant silicone pressure-sensitive adhesive coatings.

Silicone elastomer insulated cables and wiring provide connections which retain their integrity and flexibility over a wide temperature range, e.g. from −55°C or even lower to +250°C. They are fire resistant, but if damaged or destroyed by fire leave an insulating residue of silica and do not evolve highly toxic fumes.[20]

Keypads for telephones and micro-computers are now also important end applications for silicone rubber. A number of keypads can be produced by a single operation in multi-cavity moulds. This eliminates the need for a much larger number of mechanical components and assembly. The silicone elastomer used for the microcomputer keypad shown in Fig. 3.12 retains its elastic properties over a long life without the adverse change due to ageing characteristics of other elastomers. When the keys are depressed, they actuate a membrane switching sandwich below and then spring back into place. The resilience of the keypad shown was tested over five million operational cycles before series production.

Fig. 3.12. Microcomputer keypad made from a single moulding of a liquid silicone rubber, with its underlying membrane switching circuit. (Sinclair Research, Haffenden and NFI Electronics.)

3.4.7. Silicone Aids in Electronic Production

Apart from direct use as insulating and constructional materials, silicones are also used as processing aids in the production of electronic equipment.

Silicone vacuum pump fluids are used in the diffusion pumps which provide high vacua for the production of semiconductors and cathode ray tubes or other vacuum devices.

Many electronic components are encapsulated in epoxy or other hard organic coating resins. Here silicones find wide use as release agents, either as fluids or to make elastomeric moulds. Figure 3.13 shows a silicone elastomer mould used in this way for capacitor production. The silicone elastomer, a two-component room-temperature curing product, is first poured around the master prototype and sets to form an extremely accurate profile. High elongation and good values of tensile and tear strength allow the mould to be used repeatedly for long production runs.

Fig. 3.13. A mould manufactured from low consistency (RTV) silicone elastomer is used in the epoxy encapsulation of capacitors. (Replica Materials Ltd and Suflex Ltd.)

3.5. FUTURE DEVELOPMENTS

The trend towards further miniaturisation and surface mounting of components in electronic construction is likely to lead to increased use and development of high-purity silicones as coatings to provide protection against the environment. Flexible silicone adhesives will also play an important role in bonding components to substrates where differing thermal expansion coefficients preclude the use of rigid materials.

Silicone encapsulants should continue to be recognised for their benefits of easy application, fire resistance and low toxicity. They are likely to play an important role in such applications as high-voltage cascades and automotive electronic modules.

In the general field of constructional and insulation materials silicone rubber keypads are already establishing an important position and will grow in importance.

One interesting area of chemistry which may grow in importance concerns silicone–organic copolymers in which the silicone polymer chain is interconnected with large polymeric groups such as epoxy or polyester. However, while such copolymers have been successful in other application areas, for example silicone–polyester resins as paint coatings, it must be admitted that so far they have not achieved comparable importance in the field of electronic materials. This may change with further developments in chemistry, and the continued evolution of electronic packaging and construction.

REFERENCES

1. Kipping, F. S. and Lloyd, L. L., Organic derivatives of silicon, *J. Chem. Soc.*, **79** (1901) 449.
2. Kirk, R. E. and Othmer, D. F. (Eds), *Encyclopedia of Chemical Technology*, 3rd edn, **20**, 1982, 922–62.
3. Eaborn, C., *Organosilicon Compounds*, Butterworth Scientific Publications, London, 1960.
4. Noll, W., *Chemie und Technologie der Silicone*, Verlag Chemie, Weinheim, 1960.
5. Von Hippel, A., *Dielectric Materials and Applications*, Technology Press of MIT and John Wiley & Sons Inc., New York, 1954.
6. Lynch, W., *Handbook of Silicone Rubber Fabrication*, Van Nostrand Rheinhold Co., New York, 1978.
7. Lindborg, W., *Optical Fibres*, Ericsson Review, No. 3, 1980, 8–13.
8. France, P. W., Dunn, P. L. and Reeve, M. H., Plastic coating of glass fibers and its influence on strength, *Fiber and Integrated Optics*, **2** (1979).
9. Davis, J., Protection au silicone des composants électroniques automobiles, *Proceedings of 2nd. International Congress on Automotive Electronics*, Paris, 1981.
10. Davis, J., Protection of automotive electronic components and modules by silicone encapsulants, *Proceedings of International Symposium on Automotive Technology and Automation*, Stockholm, 1981.
11. Day, A. G. and Weller, M. G., Flammability testing of printed wiring board material and the effect of the conformal coating, *Proceedings of Internepcon Conference*, Brighton, 1980.
12. Davis, J. and Jones, G. M., The role of silicone transfer moulding compounds in the packaging of semiconductors and passive electronic components, *Proceedings of International Macroelectronic Conference*, Electronica, Munich, 1982.
13. Harrison, J. C., Long term reliability of plastic encapsulated devices through control of the encapsulation material, *Proceedings of Symposium on Plastic Encapsulated Semiconductor Components*, Malvern, 1976.

14. Lawson, R. W., The accelerated testing of plastic encapsulated semiconductor components, *Proceedings of the IEEE Reliability Physics Symposium*, 1974.
15. Roberts, B. C., Plastic encapsulation of semiconductor devices, *Proceedings of the International Conference Plastics in Telecommunications III*, IEEE, London, 1981.
16. Waldern, A. M., Silicones: a new approach to hybrid assembly, *Proceedings of Internepcon Conference*, Brighton, 1983.
17. Soos, N. A. and Jaffe, D., Encapsulation of large beam leaded devices, *Proceedings of the 28th Electronics Components Conference*, Anaheim, California, 1978.
18. White, M. L., Serpellio, J. W., Striny, K. M. and Rosenzweig, W., The use of silicone R.T.V. rubber for alpha particle protection on silicon integrated circuits, *Proceedings of the Reliability Physics Symposium*, Orlando, Florida, 1981.
19. Cush, R. J., LSR: the versatile alternative, *British Plastics and Rubber* (Oct. 1982) 29–31.
20. Lipowitz, J., *Combustion, Flammability and Fire Hazard Properties of Silicones*, Publication NMAB-342, National Academy of Sciences, Washington, D.C., 1978.

Chapter 4

EPOXIDE RESINS AND THEIR FORMULATION

MARTIN T. GOOSEY

Dynachem Corporation, Tustin, California, USA

4.1. INTRODUCTION

The epoxide resins are a class of materials that possess all the properties required to make them useful for a wide range of applications throughout the electronics industry. They have good electrical properties, low shrinkage, good adhesion and resistance to thermal and mechanical shock, whilst also possessing resistance to moisture, solvents and general chemical attack.

Epoxide resins are all based on the epoxide group, a strained three-membered carbon, carbon, oxygen ring structure (also known as the oxirane group)

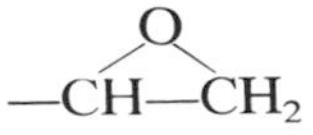

$$\overset{\displaystyle O}{\overset{\displaystyle \diagup\!\!\diagdown}{-CH-CH_2}}$$

The first synthesis of an epoxide group was made early in the nineteenth century, but it was not until 100 years later that the possible uses of epoxides began to be realised. They were first brought to the attention of industry in 1939, with the publication of a German patent by I. G. Farben[1] describing liquid epoxides. Widespread usage of epoxides was soon envisaged and this important new process was exploited by Ceiba Ltd, who introduced a class of thermosetting resins to industry at the Swiss Industries Fair of 1946. Since then many new applications have been found and the dramatic increase in their usage is illustrated by the production figures for the USA, where production rose from virtually zero in 1948 to 30 million pounds in 1957 and 130

99

million pounds in 1968. By 1978 American production had risen to 280 million pounds and the world usage of epoxides for semiconductor encapsulation alone was estimated to be well over 40 million pounds in 1983.

One of the most widely used series of epoxide resins is made by reacting epichlorohydrin with bisphenol A, in the presence of an alkali, to give the diglycidyl ether of bisphenol A (DGEBA)

$$HO-\langle\bigcirc\rangle-\underset{\underset{CH_3}{|}}{\overset{\overset{CH_3}{|}}{C}}-\langle\bigcirc\rangle-OH \; + \; 2CH_2-CH-CH_2Cl \longrightarrow$$

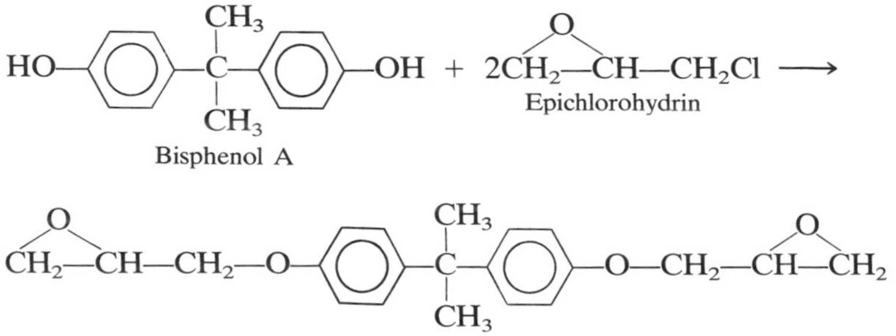

Bisphenol A   Epichlorohydrin

$$CH_2-CH-CH_2-O-\langle\bigcirc\rangle-\underset{\underset{CH_3}{|}}{\overset{\overset{CH_3}{|}}{C}}-\langle\bigcirc\rangle-O-CH_2-CH-CH_2$$

Diglycidylether of bisphenol A (DGEBA)

Bisphenol A is easily and inexpensively formed from the reaction between acetone and phenol, whilst epichlorohydrin is derived from propane. The diepoxide formed from these reactions has a molecular weight of 340 and many of the well-known commercially used resins have average molecular weights in this region. Higher molecular weight resins can be obtained by reducing the amount of epichlorohydrin used in the preparation and by increasing the pH, for example

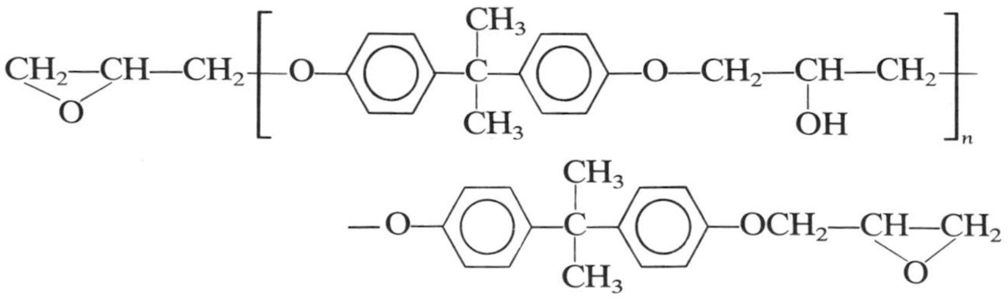

$$CH_2-CH-CH_2-\left[O-\langle\bigcirc\rangle-\underset{\underset{CH_3}{|}}{\overset{\overset{CH_3}{|}}{C}}-\langle\bigcirc\rangle-O-CH_2-CH-CH_2-\right]_n$$
$$-O-\langle\bigcirc\rangle-\underset{\underset{CH_3}{|}}{\overset{\overset{CH_3}{|}}{C}}-\langle\bigcirc\rangle-OCH_2-CH-CH_2$$

For liquid epoxy resins *n* is generally less than 1, while for the solid types *n* is generally 2 or larger, although it should be remembered that a product will always be a mixture of various molecular weight resins. As their average molecular weight increases, the resins become more and more viscous until they eventually turn into solids with increasing melting points (Table 4.1).

Table 4.1
Melting Points of Typical Epoxide Resins

Molecular weight	Average value of n/molecule	Melting point (°C)
360– 380	0·1	Liquid at room temperature
380– 420	0·2	Liquid at room temperature
460– 560	0·6	Liquid at room temperature
850–1 100	2·2	65– 75
1 750–2 050	5·5	95–105
4 000–5 000	14·4	125–135

In order to form the familiar crosslinked networks with the desired physical properties, epoxide resins are cured by the addition of a suitable chemical known as a hardener or curing agent. The term 'curing' is used to describe the process in which an epoxide and a curing agent react to give a highly crosslinked network. The curing agents for epoxides are numerous and come from well over 100 different classes of chemical compound.

The curing of epoxides occurs by an addition-type reaction, with the interaction of resin and hardener or by a catalytically induced homopolymerisation to form a crosslinked structure. These types of curing reaction give epoxides a particular advantage over some other polymers in which a condensation-type cure leads to the evolution of a small unwanted molecule such as water or an alcohol. Condensation molecules are an obvious potential source of problems when polymerisation occurs in the proximity of an electronic circuit, since they can lead to device malfunctioning or corrosion of metallisation. Another advantage offered by the addition-type curing reaction is reduced shrinkage, and epoxides are well known to have unusually low shrinkage values (usually less than 2%).

In order to be useful, a cured epoxide compound should possess as many of the following properties as possible:

(1) good adhesion
(2) environmental and chemical stability
(3) thermal stability
(4) long pot and storage lives
(5) be easily modifiable
(6) good reproducibility
(7) mechanical stability

In order to impart these properties to the final product, very careful selection of the epoxide resin and hardener system, as well as all the other components of the formulation, is vitally important. As the cured material is crosslinked, the structure and properties of the original resin will be completely changed. The material will no longer be soluble without decomposition and its chemical resistance will depend as much upon the hardener selected as upon the epoxy resin. As an illustration of the dependency of the cured system upon the hardener, acid anhydride based hardeners offer excellent resistance to acid environments, whilst amine-type hardeners offer maximum resistance to alkaline environments.

The reaction between an epoxide resin and a hardener, although usually leading to a solid material, is unlikely to be sufficient to impart all of the desired properties. In order to obtain the characteristics that will make it useful for a specific application, a variety of other compounds may be added to the resin–hardener system during the formulation stages. For the sake of formulation purposes, an epoxide resin system can be considered to be comprised of two principal components known as the binder and the filler. The binder component of the system contains the reactive ingredients and is the part that undergoes the chemical reaction when exposed to heat or possibly light and pressure. Ultimately the binder component is responsible for providing cured properties compatible with the particular application the material is used for.

The filler component is comprised of all the other materials that are needed to complete the formulation. These are generally non-reactive, although two particular exceptions are the silane coupling agents used as reactive surface coatings and reactive diluents. Many different compounds find application as filler components but the more generally used ones are extenders such as silica, alumina, talc and other minerals, pigments to give the desired coloration and, if necessary, release agents to prevent sticking after moulding. The composition of a typical epoxide transfer moulding compound used for semiconductor device encapsulation is shown in Table 4.2. As can be seen from this formulation the constituents comprising the filler portion are very varied. In addition to the inorganic extenders, other organic materials are added to improve the properties. Quite often the only unmodified epoxide resins suitable for a particular application may have certain disadvantages, such as high cost, high viscosity or excessive rigidity. The resin may then be modified by adding compounds known as

Table 4.2
Composition of a Typical Semiconductor Grade Epoxide Transfer Moulding
Compound

Binder component	Filler component
Epoxy cresol novolac	Silica (crystalline or fused)
Cresol novolac	Silane coupling agent
Accelerator	Pigment (carbon black)
Flame retardant epoxy	Flame retardant (antimony oxide)
	Release agent (wax)

diluents or flexibilisers and it may even be blended with other resins. Materials may also be added in order to counteract the effects of any deleterious impurities in the compound.

Once it is appreciated that a typical epoxide compound may be comprised of many individual ingredients, all of which are available with a variety of chemical and physical forms, it is not surprising that, before a compound can be successfully formulated, some knowledge of the likely application is required. The major considerations that will lead to an initial selection of materials are:

(1) What object is the epoxide to be used with?
(2) What are the cost limitations?
(3) What environment is the object to be exposed to?
(4) How is the epoxide to be applied to the object?

Having decided upon these factors there will then be a set of considerations which will apply with every application.

(1) What shelf-life must the material have?
(2) What physical properties must the cured material have?
(3) What cure time and temperature are required?
(4) What viscosity must the epoxide have?
(5) What chemical and thermal stabilities are required?
(6) Is good adhesion required?
(7) Does the material need to release easily from a mould?
(8) Does the material have to be especially pure?

Quite often there will be a large number of contrasting requirements in order to make the material with the desired properties and it will be an unusual case where all of the desired characteristics can be accommodated. For instance, a system that gives the desired rapid cure may

104 Martin T. Goosey

not have the correct storage lifetime. Alternatively a system that provides the correct shelf stability may need an expensive latent catalyst system which exceeds the permitted cost limitations for the compound.

Not only is there a large degree of latitude in the selection of individual ingredients for an epoxy resin formulation but there is also a degree of flexibility in the final physical form of the uncured compound. This will obviously be determined by the application, but an epoxy formulation may be required as a powder, a liquid or a two-part system which reacts upon mixing. If a single-part material is required, whether powder or liquid, the individual components are mixed thoroughly together either in solution using suitable solvents or by melt blending. In the melt blending process heat is used to melt the solid ingredients so that they are thoroughly mixed together and so that the unfusible inorganic constituents are completely coated and evenly dispersed within the binder materials. This process obviously has to be carried out at a temperature high enough to fuse the organic components but also at a temperature low enough to avoid excessive curing of the binder system. The solvent process simply involves dissolving the binder system in one or more suitable solvents and then adding the fillers. The solvents can be carefully removed under elevated temperature and reduced pressure conditions. In this process the binder must be able to withstand the temperature of solvent removal if a dry product is required. Two other processes are also commonly used and these are the liquid process and the dry process. In the liquid process one or more of the ingredients is a liquid at room temperature and the final product will be a solvent-free liquid, although partial curing may be carried out to solidify the material for a dry product. The dry process simply involves mixing finely divided powders of the solid ingredients. The process may begin with large particle size ingredients, but during the mixing the total mix is reduced in particle size yielding a fine homogeneously blended powder.

In order to obtain an epoxide compound suitable for a particular application, all the above considerations have to be taken into account and the successful formulation can only be made with full knowledge of the intended application. Each of the major components commonly used in epoxy resin formulations is now individually discussed and the influence of physical and chemical properties on the properties of the final formulation described.

4.2. EPOXIDE RESINS

The epoxide resins, whilst being characterised by the epoxy group, are available with greatly varying physical and chemical properties. They exhibit widely ranging molecular structures, molecular weights and viscosities. This wide range of epoxide resins means that they find broad usage in all types of application, particularly in the electronics industry. By careful selection of the correct resins and other components of the formulation the epoxy resins can be made to yield a variety of forms, ranging from soft and flexible materials to hard, tough, highly crosslinked, chemically resistant products.

One of the most commonly used groups of epoxide resins is that comprised of the diglycidyl ethers, and these are made by reacting epichlorohydrin with a bisphenol such as bisphenol A. The preparation of the diglycidyl ether of bisphenol A has already been illustrated, but many other glycidyl ether type epoxides are of considerable importance. The preparation of an epoxide novolac type resin is shown below:

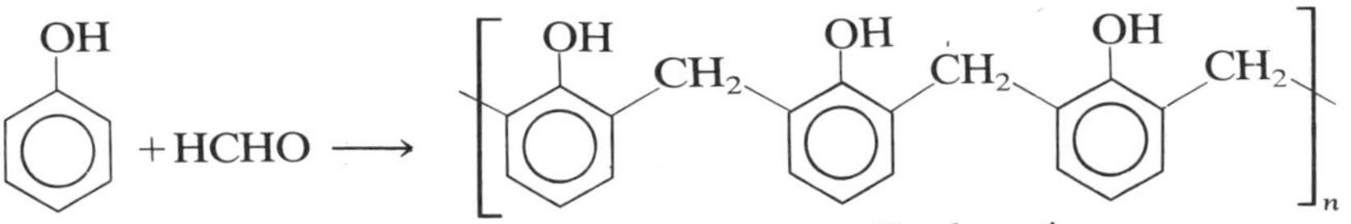

Novolac resin   (1)

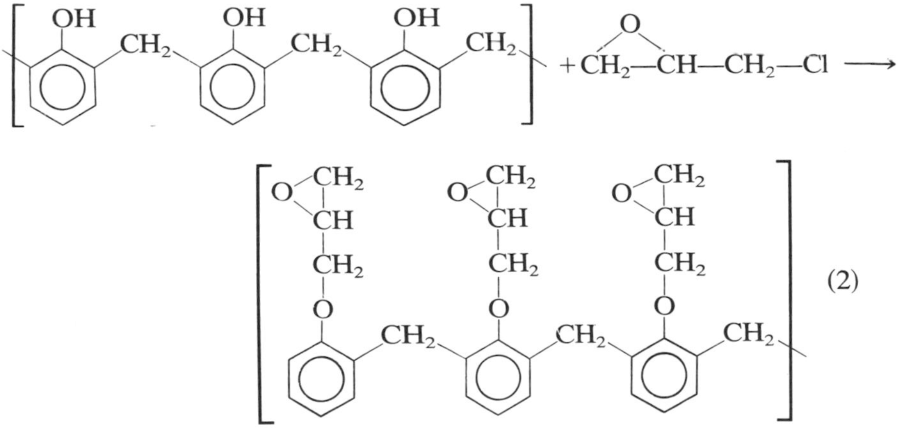

Epoxide novolac   (2)

The epoxide novolacs are a very important class of compounds, since they are the basis of the materials used in the formulation of epoxide moulding compounds. They are ideal in these applications because very high crosslink densities are achieved in the cured polymer network. This is a direct result of the large number of epoxy groups available in the resin for reaction and these also give the added advantage of high temperature stability. A typical epoxy novolac might have a molecular weight of around 650, with a weight per epoxide of 175–182 and a very high viscosity. In order to develop high temperature stability and the other properties characteristic of the novolac epoxide, a high temperature cure is necessary, usually in the region of 160–180°C. When an epoxide novolac is used in a room-temperature curing system the properties are likely to be similar to those of a bisphenol A based material in the same application.

In addition to the diglycidyl ether based resins, there are also non-glycidyl ether based epoxide resins. These materials can usually be prepared by carrying out an epoxidisation reaction on an unsaturated compound using hydrogen peroxide or peracetic acid:

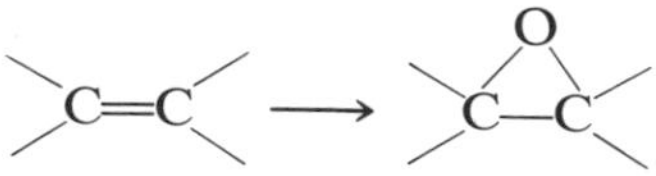

$$\text{C=C} \longrightarrow \underset{\displaystyle \text{C}-\text{C}}{\overset{\displaystyle O}{\triangle}}$$

These are also available in a variety of forms, but can essentially be divided into those which contain a ring structure as well as an epoxide group (cycloaliphatic) and those which have a linear structure attached to the epoxide group (acyclic aliphatic).

The cycloaliphatic resins are a paler colour than the standard diglycidyl ether resins and also have much lower viscosities. Typical examples are dicyclopentadiene dioxide, which is a solid at room temperature, and vinyl cyclohexene dioxide, which is a liquid

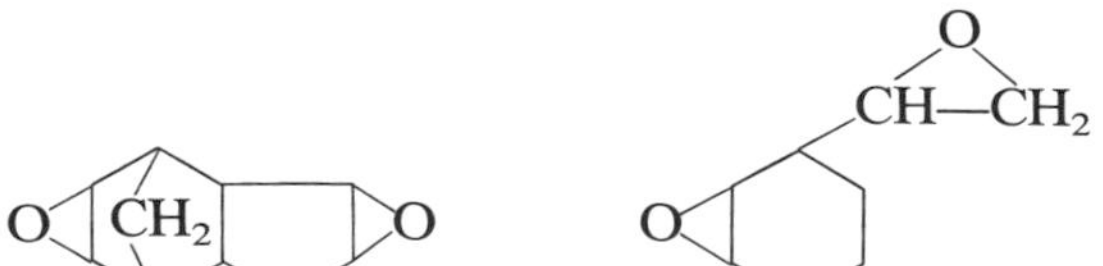

Dicyclopentadiene dioxide   Vinyl cyclohexene dioxide

They have very compact structures which lead to a much greater crosslinking density than occurs with the standard diglycidyl ethers. The lack of flexibility, which can be attributed to the absence of the oxygen containing ether linkage, also leads to a more rigid crosslinked structure. The fully cured cycloaliphatic epoxy resins thus give very high glass transition temperatures, lower dissipation factors, good arc resistance and high temperature stability. Their rigid structures also make them unsuitable for encapsulation and potting purposes. They are, however, successfully used in injection moulding and extrusion techniques and are also useful diluents for the standard diglycidyl ether resins.

Burhans[2] reported a high performance cycloaliphatic epoxide system for electrical apparatus which provided a unique balance of high heat distortion temperature, thermal shock resistance and high temperature electrical properties. All of the formulations were based upon the cycloaliphatic epoxide:

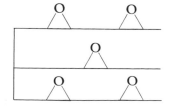

In both filled and unfilled forms they were ideally suited to the encapsulation of coils, transformers and other electrical components. The use of a dual catalyst system permitted rapid curing at low temperatures.

The acyclic aliphatic resins differ from the cyclic aliphatic epoxy resins in that they do not possess ring structures. They can be sub-divided into two groups: the epoxidised diene polymers and the epoxidised oils. The epoxidised diene resins are generally more viscous than the diglycidyl ethers, except when volatile diluents are used. They are also less reactive with amines than diglycidyl ethers because of the absence of the ether linkage, although they have similar reactivity with anhydride hardeners. The epoxidised oils have the general skeletal structure:

The number of epoxide groups per molecule can vary, although an average four to six groups would be usual. As with the epoxidised dienes, their reactivity with amine curing agents is also low because of the absent ether linkage. The epoxidised oils are usually modified before use by the introduction of hydroxyl groups into the molecule in order to reduce the spacing between potential crosslinking sites. By introducing these hydroxyl groups, the modified material can be made to exhibit increased toughness and rigidity.

4.3. CURING AGENTS

The special properties that have allowed such widespread usage of epoxide resins are only obtained when the resins are properly cured. The final cured characteristics of an epoxide resin are largely dependent upon the curing system that is used to convert the resin into a thermoset product. This reaction is brought about by a compound generally known as a hardener, which causes the epoxide resin to undergo either an addition or a catalytic reaction. In the case of an addition reaction, the hardener molecule is chemically combined into the polymer matrix serving as a bridge for crosslinking several resin molecules. This is a type of polymerisation reaction referred to as heteropolymerisation. In the catalytic type of reaction, the hardener behaves as a catalyst, encouraging the self-polymerisation of the resin and this is known as homopolymerisation.

There are many different types of hardener and catalyst available and the curing agent will largely determine the storage life, the reaction rate, the time and temperature required for curing, and the final properties of the system. All of the various hardeners and catalysts have distinct advantages and limitations associated with them and this section details some of the major curing agents commonly used.

When considering the formulation of an epoxide resin system for a particular application the final selection of a curing agent will depend upon several important factors such as:

(1) The physical, electrical, chemical, thermal and mechanical properties required of the cured material.
(2) The limitations on cure, post-cure and temperature.
(3) The cost of the curing agent.
(4) The limitations on the material concerned, e.g. pot life, curing exotherm, viscosity and toxicity.

Although there are literally hundreds of curing agents to choose from, the commonly used ones can be conveniently divided into the major categories aliphatic amines, aromatic amines, acid anhydrides and catalytic curing agents or latent hardeners.

4.3.1. Aliphatic Amines

The aliphatic amines form a group of low viscosity, low cost, room-temperature curing agents which are probably the most widely used of all epoxide hardeners. They can be used unmodified and also in a reacted form which reduces their reactivity and the curing exotherm. Too rapid a cure reaction can complicate any attempt to use the materials for a production process because the resin and hardener have to be mixed prior to use. A large curing exotherm is also undesirable because it limits the casting of large pieceparts where thermal runaway might occur.

The aliphatic polyamines most commonly used to cure epoxides are diethylamine triamine (DETA) and triethylene tetramine (TETA). These are both pungent, low viscosity liquids and they are commonly used with the diglycidyl ether of bisphenol A type resins, where rapid low temperature cures arc required. Three other commonly used polyamines are ethylene diamine (EDA), tetramethylene pentamine (TEPA) and diethylaminopropylamine (DEAPA)

$$H_2NCH_2CH_2NHCH_2CH_2NH_2 \qquad H_2NCH_2CH_2NH_2$$

Diethylene triamine (DETA)   Ethylene diamine (EDA)

$$H_2N[(CH_2)_2NH]_2(CH_2)_2NH_2 \qquad H_2N[(CH_2)_2NH]_3(CH_2)_2NH_2$$

Triethylene tetramine (TETA)   Tetraethylene pentamine (TEPA)

$$(CH_3CH_2)_2N(CH_2)_3NH_2$$

Diethylaminopropylamine (DEAPA)

The properties of these amines are given in Table 4.3.

Epoxides cured with DETA and TETA usually have similar properties. They have good chemical resistance to aqueous acids and alkalis, although they are attacked by strong organic acids and nitric acid. These properties are maintained up to moderate temperatures, but for high temperature performance an aromatic amine would be preferred.

DEAPA is very similar to DETA, yet it offers pot lives of up to 4 h at room temperature and needs to be heated in order for full curing to be achieved. It does not crosslink so densely as the TETA-type amines and thus its chemical properties are somewhat inferior.

Table 4.3
Properties of Commonly Used Polyamines and Optimised Amounts Required
to Cure Epoxides

Amine	Molecular weight	Active hydrogens	Calculated amount (phr)	State
DETA	103	5	11	Liquid
TETA	146	6	13	Liquid
TEPA	189	7	14	Liquid
EDA	60	4	8	Liquid
DEAPA	130	2	7	Liquid

Epoxide polyamine systems are commonly used as the basis for solvent-type epoxide coatings. They offer excellent coverage of both ferrous and non-ferrous metals, as well as concrete and wood. These coatings are usually applied by conventional spraying or brushing techniques. They are also often used as dip coatings for various electronic components and electrical equipment, where the epoxide resins will be based on DGEBA or a low molecular weight epoxide novolac. The formulated liquid system may then be modified with a thixotroping filler or filler combination to give regular coating and non-drip characteristics. A typical basis for a formulation would be:

Material	Quantity (phr)
DGEBA	100
Bentonite	16
Iron oxide	4
DETA	10

Several other types of curing agent based upon the aliphatic amine approach are also commonly used in epoxide formulations. Examples are the glycidyl adducts of the aliphatic amines, in which the aliphatic amine is partially reacted with an epoxide to give an adduct with low volatility. Similar types of adduct can also be made by reacting the amines with ethylene and propylene oxides. By forming such adducts, reaction rates can be increased, but the physical properties may deteriorate somewhat, and, in the case of the ethylene and propylene oxide adducts, the compounds are hygroscopic.

Another group of curing agents based upon the polyamine concept is the cycloaliphatic amines. These have properties somewhere between those of the aliphatic amines and the aromatic amines. Typical examples are aminoethyl piperazine and methane diamine

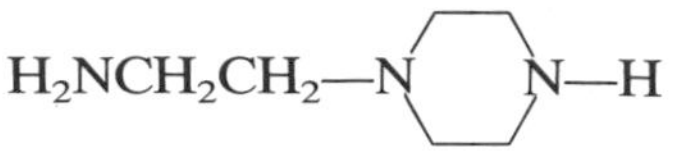

Aminoethyl piperazine 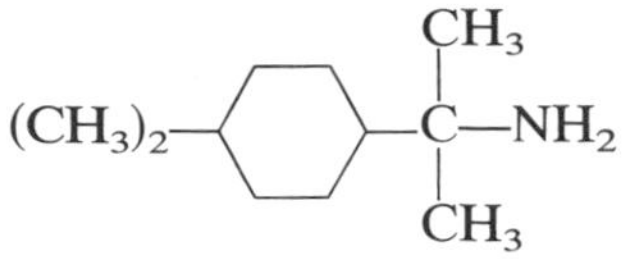   Methane diamine

These materials generally give higher glass transition temperatures than the aliphatic amines but have poorer thermal stability than the aromatic amines.

4.3.2. Aromatic Amines

The aromatic amines are characterised by having the nitrogen atoms of their amino groups attached directly to an aromatic ring structure. They offer several advantages over the aliphatic amines, such as higher glass transition temperatures, longer pot lives and greater chemical resistance after cure. Their reactivity is generally much less than that of the aliphatic amines, with cures often taking several months at room temperature. As a general guideline, the reaction rates for aromatic amines at around 90°C are roughly comparable to the room temperature reaction rates achieved with the aliphatic primary amines.

One of the most commonly used aromatic amines is metaphenylene diamine (MPDA), a solid with a melting point around 63°C

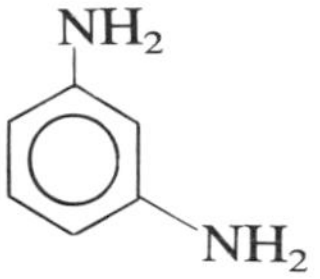

As can be seen, MPDA contains four active hydrogens and the stoichiometric amount required to cure DGEBA is approximately 14·5 phr. In order to thoroughly mix the resin and hardener, it is usual to heat about 15% of the resin to 65°C and then add the molten MPDA. After allowing the mixture to cool to room temperature, the rest of the resin can then be added. When used in the stoichiometric quantity, a 500 g batch would have a pot life of around 2–3 h at 50°C.

Two other important aromatic amine curing agents are 4,4′-diaminodiphenylmethane (DADM) and 4,4′-diaminodiphenyl sulphone (DADS)

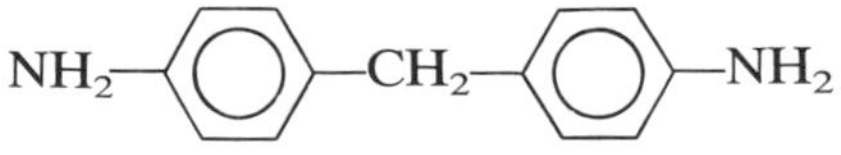

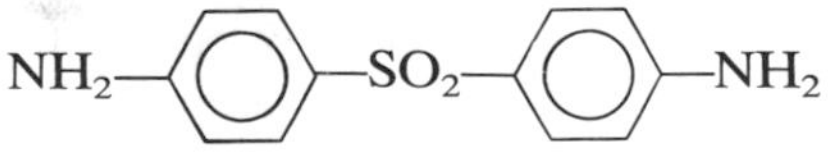

4,4′-Diaminodiphenyl methane   4,4′-Diaminodiphenyl sulphone
(DADM)   (DADS)

Martin T. Goosey

DADM is a solid, melting at around 90°C. It contains four active hydrogens and is usually used in concentrations of 25–30 parts per hundred of resin. Mixing is achieved in the same way as with metaphenylene diamine. The resin is heated to just above 90°C and the molten amine added. After thorough mixing, the batch is cooled by the addition of more resin and mixing is continued. DADM does not give the high temperature strength of MPDA, but its electrical properties, such as dielectric constant and dissipation factors, are enhanced.

DADS has an advantage over other aromatic amines in that it gives very high glass transition temperatures and its strength properties are maintained to higher temperatures. It is a solid with a melting point around 175°C and is generally used in the range of 40 phr with DGEBA type epoxides (although 20–30 phr may be adequate if an accelerator is utilised). In order to mix the resin and hardener, the resin is heated to above 130°C and the DADS is added slowly with stirring and maintenance of temperature until all the hardener is dissolved. If an accelerator is to be used, the temperature should be lowered to 100°C and one part by weight of boron trifluoride monoethylamine blended in. This type of system would be suitable for electrical castings requiring extreme service durability and elevated temperature strength.

As well as the three principal aromatic amines reported above, various other aromatic amines find use in certain applications. Typical examples of these compounds are

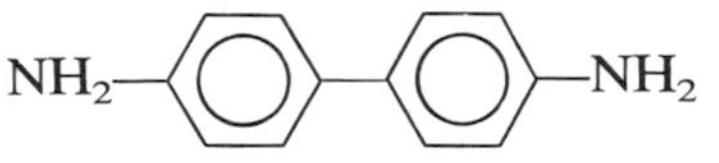

Benzidine

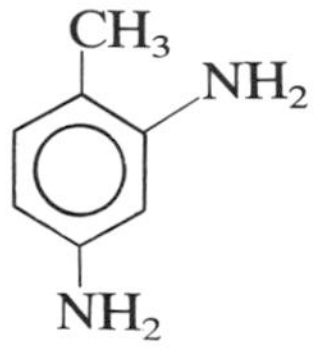

Toluene diamine

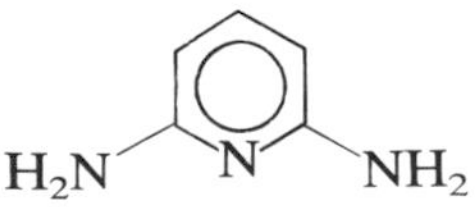

2,6-Diaminopyridine

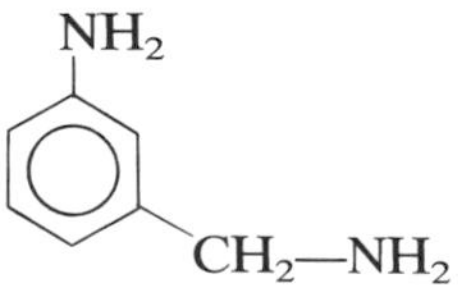

1-Aminobenzylamine

Similarly to the aliphatic amines, the aromatic amines may also be adducted with a variety of compounds in order to alter their properties. Eutectic blends of certain aromatic amines have also been made which give the advantage of allowing mixing to be carried out at room temperature.

4.3.3. Anhydrides

Acid anhydrides are the second most commonly used curing agents, and they are characterised by the anhydride group, which is derived from a diacid by dehydration

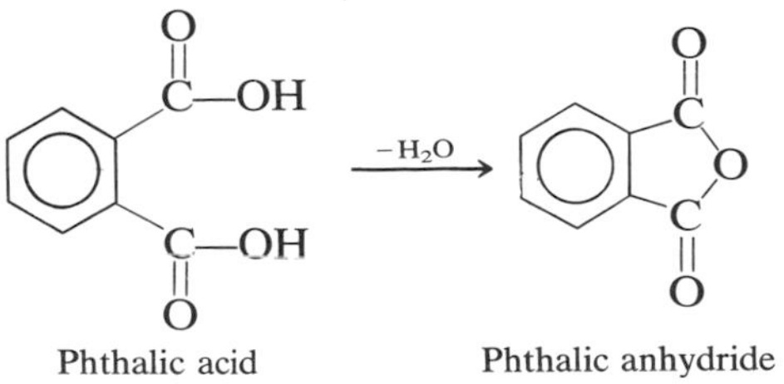

Phthalic acid   Phthalic anhydride

They represent a group of hardeners which are low cost, provide long pot lives, have excellent heat stability and are especially suited to electrical applications. Anhydrides also offer advantages in handling, since, unlike the amines, they are not skin-sensitising agents. Also, unlike the aliphatic amine cured epoxide systems, the acid anhydride cured epoxides usually require long, high temperature cure schedules before optimum properties can be achieved. This is probably the major limitation to their use, although certain accelerators are useful in enhancing the reactivity. Acid anhydrides can, however, provide more rapid cures than amines when used as curing agents for the cycloaliphatic and epoxidised olefin-based resins.

The curing reactions of anhydrides with epoxide resins are quite complex and may be dependent upon a number of factors, such as the temperature, the type of epoxide, the amount of anhydride used and the presence and type of accelerator. The first stage of the curing reaction is the opening of the epoxide ring by a hydroxyl group, such as may be found in the epoxide resin

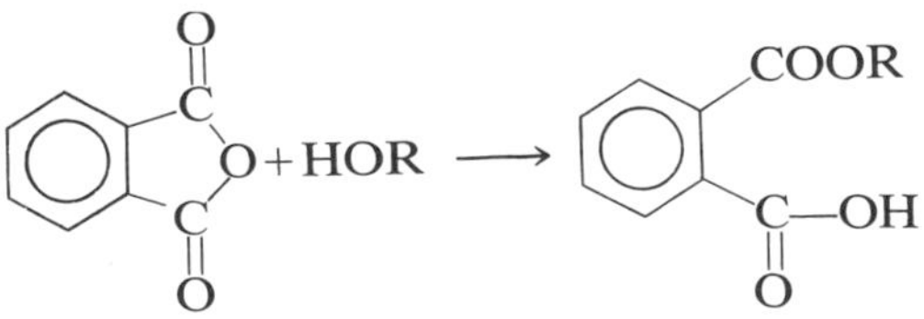

There are then five further possible types of reaction that can take place but the two most important ones are:

(1) the reaction of a carboxylic acid group with an epoxide group

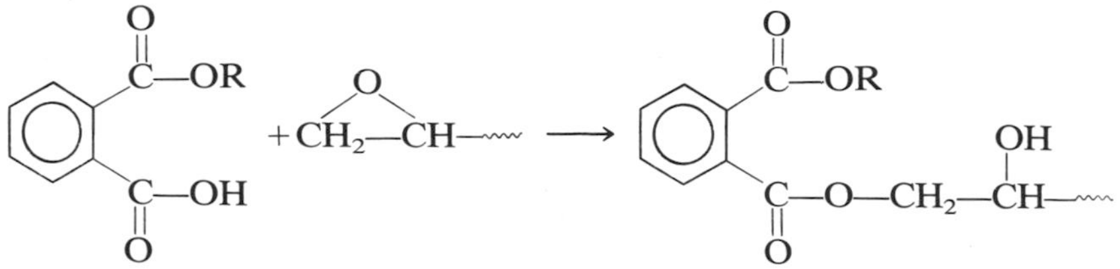

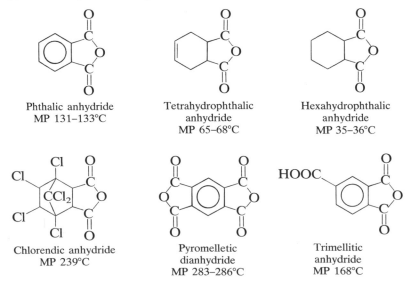

(2) the etherification of an epoxide group by a hydroxyl group

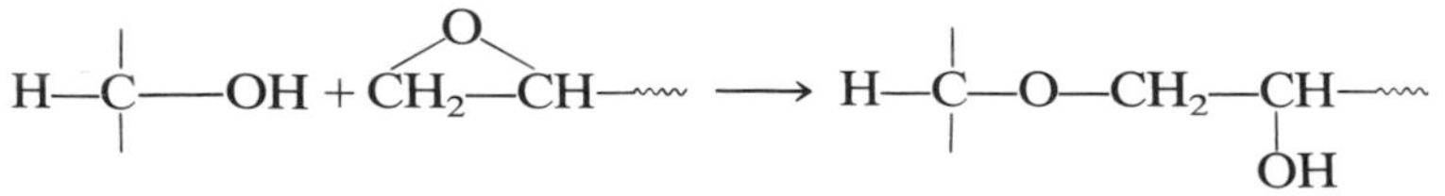

During the curing reaction it is quite often found that approximately equal amounts of the ester and ether linkages are formed, although the reaction is usually modified in commercial systems by the use of accelerators such as tertiary amines.

There are many types of anhydride available, but one of the most important groups is that based upon anhydrides with cyclic structures. Typical examples of anhydrides in common use are

Phthalic anhydride
MP 131–133°C

Tetrahydrophthalic
anhydride
MP 65–68°C

Hexahydrophthalic
anhydride
MP 35–36°C

Chlorendic anhydride
MP 239°C

Pyromelletic
dianhydride
MP 283–286°C

Trimellitic
anhydride
MP 168°C

The tertiary amine accelerators are commonly used with anhydride systems to speed up their curing reactions. They are used in concentrations of 0·5 to 5 phr and typical examples are

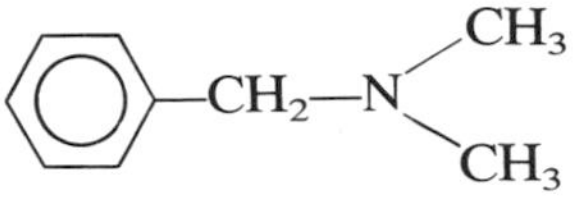

Benzyl dimethylamine

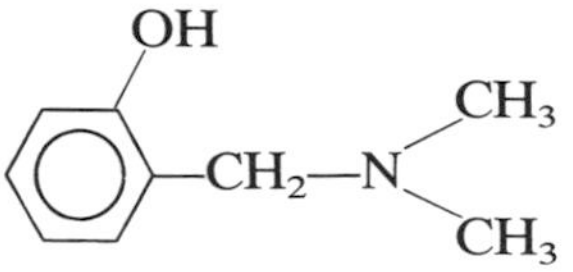

Dimethyl aminomethylphenol (DMP 10)

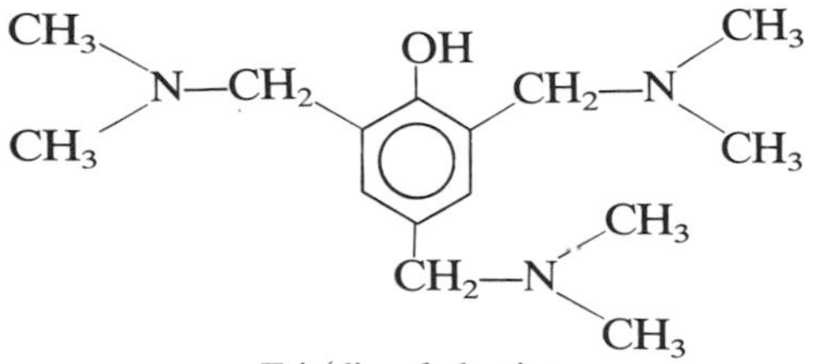

Tris(dimethylamino-methyl) phenol (DMP 30)

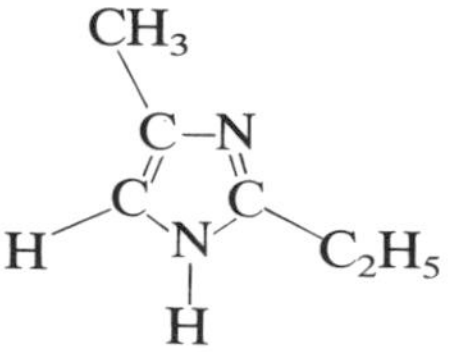

2 Ethyl 4 methylimidazole

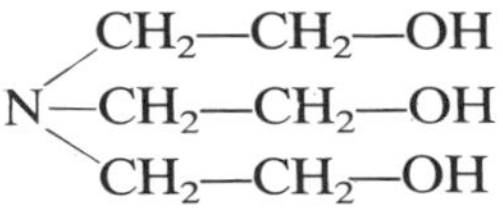

Triethanolamine

Phthalic anhydride is used extensively in low cost epoxy formulations since it is itself very inexpensive. Its main application is with glycidyl ether type resins, in the production of large electrical castings. The anhydride is usually mixed with the resin by stirring it into the hot resin at 120–140°C and then cooling to 60°C to prolong the pot life. With low molecular weight bisphenol A resins, phthalic anhydride is usually used at a concentration of between 30 and 45 parts per hundred of resin. The curing reaction is fairly slow, examples being 24 h at 120°C, 16 h at 130°C and 6 h at 150°C. Phthalic anhydride cured systems do not give particularly high glass transition temperatures and problems may be experienced with sublimation of the anhydride. The main advantages of phthalic anhydride cured systems are the very low curing exotherms, which allow the production of large castings with good chemical resistance and good electrical properties.

Tetrahydrophthalic anhydride is often used in place of phthalic anhydride, since it is of comparable cost and has the advantage of not subliming. Even more useful is hexahydrophthalic anhydride, which imparts good all-round properties and is also soluble in epoxide resins at room temperature. It is used in concentrations of between 55 and 80 parts per hundred of resin and, in the presence of around 0·5% of benzyl dimethyl amine, it gives similar cure times to phthalic anhydride, with slightly higher glass transition temperatures.

Where higher glass transition temperatures are required with diglycidyl ether of bisphenol A based resins, anhydrides with higher functionality have been employed, such as pyromellitic dianhydride, which contains two anhydride groups, and trimelletic anhydride, which has an additional carboxylic acid group. Chlorendic anhydride is a member of the family of chlorinated and brominated anhydrides that, because of their halogen content, impart flame retardancy upon a cured resin. Systems cured with chlorendic anhydride have excellent electrical and mechanical properties although it does have a somewhat limited thermal stability and a tendency to be hygroscopic.

Similarly to the aromatic amines, eutectic mixtures of two anhydrides are sometimes used to overcome difficult mixing procedures such as the requirement for high temperature mixing. One such example is a 70/30 mixture of chlorendic anhydride and hexahydrophthalic anhydride which imparts nearly the same properties on the cured system as pure chlorendic anhydride yet allows much simpler mixing to be achieved.

4.3.4. Catalytic Curing Agents

The amine and anhydride curing agents described above lead to the formation of the crosslinked thermoset compound primarily by a poly addition reaction. In contrast to this, the catalytic curing agents achieve the same result by causing the opening of the strained epoxide ring, which leads to homopolymerisation. The resin's molecules react directly with each other, the catalytic curing agent not being incorporated into the reaction product. This results in the crosslinked polymer having a largely polyether-like structure.

The most popular catalytic curing agents are the Lewis acids and Lewis bases. A Lewis acid is a species that can accept an electron pair during a reaction and a Lewis base is one that can donate an electron pair. Typical Lewis acids which find use with epoxy resins include zinc chloride, boron trifluoride, titanium tetrachloride and stannic chloride. The major Lewis bases are the tertiary amines although some metal alkoxides and alkali metal hydroxides are also able to initiate epoxide resin polymerisation.

Boron trifluoride is one of the most important Lewis acids currently used. However, as it is a corrosive gas which is extremely reactive at room temperature, it is usually complexed with an amine such as monoethylamine

$$\underset{\underset{F}{|}}{\overset{\overset{F}{|}}{F—B}} : \underset{\underset{H}{|}}{\overset{\overset{H}{|}}{N}}—CH_2CH_3$$

The BF_3:MEA complex

This complex has negligible reactivity at room temperature (good latency) but at around 90°C it begins to dissociate and react with the epoxide. The complex is normally used in concentrations of around 3 parts per hundred of resin and a typical cure would be 2 h at 105°C, with a post-cure of 4 h at 175°C. The cured products are very hard, strong and brittle, and have good to average electrical properties. Care should be exercised in the use of these materials with certain metals or sensitive devices, since the liberated BF_3 may cause corrosion.

Of the Lewis bases, the compounds dimethylaminomethyl phenol, (DMP 10) and tris (dimethylaminomethyl) phenol (DMP 30) are widely used as curing catalysts. They are employed in concentrations of 4–10 parts per hundred of resin with liquid diglycidyl ethers and curing takes place rapidly even at room temperature. Secondary amines also act as Lewis base type catalysts once their active hydrogen atoms have reacted with an epoxide group. A typical example of this type of compound is 2-ethyl 4-methyl imidazole, which finds use both as a curing agent by itself and as a catalyst for anhydride systems.

4.4. EXTENDERS

The term extender is very broad and covers a wide range of materials but it generally applies to inorganic materials that are added to a polymer in relatively high concentrations. One of the major reasons for using extenders in an epoxide resin system is to reduce the cost of the final product and this has become particularly important since the onset of crude oil price increases in the early 1970s. By using extenders in a resin system, not only can the price be controlled but their use also allows the properties of the finished product to be tailored for a particular application. In the electronics industry, the use of extenders with epoxides is usually more important from the point of view of tailoring properties than of cost reduction.

Extenders can be used to control curing exotherms and moisture permeability, to impart increased strength or to tailor properties such as hardness, heat distortion, thermal expansion coefficients and thermal conductivity. More novel fillers can be used to impart electrical conductivity and magnetic properties or to reduce the compounds' density by the incorporation of small hollow glass spheres. The selection of the correct filler is of the utmost importance and all the critical desired properties of the cured compound must be taken into account before a particular filler can be decided upon.

One of the first problems that is encountered is the selection of an extender with the correct particle size, shape and particle size distribution. Since the properties a filler imparts upon a formulated epoxide resin system may well be very dependent upon surface area, the particle size of a given type of filler may be critical. For a moulding compound formulation, particle size will control the flash and bleed characteristics and, indeed, the mouldability. Another important factor that has to be acknowledged is the possibility of a reaction occurring between the extender and the resin system, and since reactivity will be influenced by surface area, this should be borne in mind when selecting a particular particle size material.

Perhaps some of the most important considerations of extenders for epoxies to be used in electronic applications are the changes that they impart to the electrical properties of the formulated product. If good insulative properties are required of the compound, the condition and type of extender selected will be important since ions, salts, moisture or water-soluble impurities may be adsorbed on the particle's surfaces, all of which can serve to increase the conductivity of the compound. Conversely, high conductivity may be required, as in the case of conductive die attach materials where the extender may be a finely divided metal such as silver or gold. Care also has to be taken when selecting organic extenders, since high voltages or areas of high current density can cause tracking or carbon bridges to form. The use of extenders can thus be seen to play an important part in determining the electrical properties of a finished product and extreme care should be exercised in their selection.

Extenders are often used to alter the thermal expansion coefficient (α) of a material and this is particularly true with electronic applications, where the principal substrate is usually silicon. If a silicon die is encapsulated in an epoxide-based moulding compound, reliability problems can occur because of the thermal expansion coefficient mismatch between the silicon and the encapsulant. During any temperature changes, the differential expansion of the two materials can cause stresses to build up in the package, which lead to wire bond fatigue and possibly even cracking of the chip. The same applies with die attach materials, particularly when the adhesive used is a fairly rigid, highly crosslinked type.

It is generally true that nearly all commercial epoxide resins have extenders incorporated in their formulation, the final selection being determined by the properties to be altered, the application and the other components of the system. Some of the properties affected by the addition of extenders are listed in Table 4.4. (For some specialised applications, certain of the listed disadvantages may in fact be advantages.)

Table 4.4
Advantages and Disadvantages of the Use of Extenders

Advantages	*Disadvantages*
Reduced cure shrinkage	Increased density
Reduced cost	Reduced flow
Reduced thermal expansion coefficient	Increased dielectric constant
Reduced curing exotherms	Loss of transparency
Improved thermal stability	Promotion of crack propagation
Reduced moisture, solvent and gas permeability	Introduction of impurities
Increased thermal conductivity	Reduced tensile strength
Increased hardness	Reduced flexural strength
Improved compressive strength	Variable machinability
Improved dielectric strength	
Improved abrasion resistance	
Improved adhesive strength	
Reduced creep and stress relaxation	
Increased heat deflection temperatures	
Enhanced flame retardancy	
Improved arc resistance	
Increased pot life	

As can be seen, for the majority of applications the advantages of using extenders far outweigh the disadvantages and it is hardly surprising that the inorganic extenders find such widespread usage. Several of the more important extenders will now be discussed in more detail.

4.4.1. Silica (SiO_2)

Silica is usually thought of quite simply as SiO_2, but in fact there are over 20 distinct phases of silica, each one still being chemically SiO_2. One of the most important types of silica for use in electronic grade epoxides is fused silica. This is produced by subjecting high purity crystalline quartz to very high temperatures in an arc furnace. It has a thermal expansion coefficient of around 9.54×10^{-6} m/m/°C over the temperature range 0–1000°C and a comparatively low density. The low density means less weight increase in an epoxide formulation and thus 1 kg of fused silica will give a larger volume of product than the same weight of a denser material. The densities of various extenders are shown in Table 4.5.

Another commonly used form of silica is quartz, this has advantages over fused silica for certain applications due to its higher thermal conductivity. Thus where enhanced thermal conductivity is required, as in the encapsulation of power devices or integrated circuits (IC's) with high circuit density, quartz would be a preferred filler.

Another popular form of silica that is particularly useful with liquid systems is fumed silica and this is an amorphous form that has a very large surface area. It can be used in liquid epoxide systems to provide thixotropic and antisettling characteristics and in moulding compounds to control flash and bleed. The thickening effect of fumed silica results from the interaction of silanol (SiOH) groups on adjoining particles, as well as between the silanol groups and the reactive groups of the hardener (such as amine groups). Fumed silica is also useful in elas-

Table 4.5
Densities of Some Widely Used Fillers

Filler	Density (g/cm^3)
Fused silica	2·15
Glass	2·70
Quartz	2·65
Calcium carbonate	2·70
Alumina	3·97

tomeric materials, such as the silicone rubbers, as a reinforcing agent to improve mechanical properties. It has the distinct advantage of allowing the formulation of silicones that are translucent or even transparent.

4.4.2. Talc

Talc is the name given to hydrated magnesium silicate, which has the theoretical formula $3MgO \cdot 4SiO_2 \cdot H_2O$ and it occurs in several ores from which talc products can be made. The form of most use in plastic formulations is finely ground talc, which consists of thin platelets. It is considered to be a reinforcing filler in many compound applications, although it can also be described as an extender because of its low cost.

Talc possesses a good degree of chemical resistance, being inert to most acids, and it is also a good insulator. This latter property makes it particularly useful in the manufacture of high-frequency electrical insulators. It has good thermal stability up to 900°C, which is considerably higher than any temperature likely to be encountered in an epoxide resin application. The use of talc in epoxide formulations leads to higher stiffness and improved creep resistance at both room and elevated temperatures. In order to obtain the full benefits from using talc, it is very important that the correct compounding techniques are used to incorporate high loading into the resin system.

The machinability of epoxides varies considerably from filler to filler, and, if ease of machinability is required, the use of talc may provide this property. Talc is also of use in epoxide resin formulations when the thermal expansion coefficient has to be lowered by a large amount, since loadings up to 80 parts per hundred of resin can be tolerated. Loading epoxides with talc has also been shown to increase their lap shear strengths when they are used as adhesives.

4.4.3. Metallic Fillers

Metallic fillers can impart special properties to an epoxide formulation, such as giving response to magnetic fields and providing a radiation shielding capability. They are also used to enhance properties such as thermal conductivity and electrical conductivity and it is this last feature that makes them so useful in the electronics industry, where they find wide use in die attach materials. The principal metals used for enhancing electrical conductivity are silver, copper and gold, although various metal alloys such as silver/palladium have also been utilised.

Generally, however, alloys have not given the reliability required and the preferred materials for military applications are gold-filled.

The conductivity of the final product may vary greatly, depending upon the metallic filler selected, the particle size, shape, previous history and the amount used. With non-noble metals, the surface of the filler particles has to be cleaned to remove non-conducting surface oxides. With certain metallic fillers these oxides can gradually re-form, reducing the conductivity as the material ages.

As already stated, the addition of metallic fillers to a formulation serves to decrease the electrical insulation, but there may be other effects on the compound's electrical properties that may need to be taken into account. The frequency dependence of the dielectric loss factor increases as the metallic particles offset the low loss factors of the binder system. The loss factor is defined as the product of the power factor and the dielectric constant and is a measure of the signal absorption by the compound. Normally, low loss factors are desirable, particularly where a material is to be used in devices operating at high speed such as gallium arsenide based semiconductors, and this should be taken into account when formulating with conductive extenders.

The dielectric strength will decrease with increasing metal loading and, at the same time, the dielectric constant will increase. As the loading increases the resistivity decreases and this may negate any advantages gained from the increased dielectric constant.

The metallic fillers or extenders are usually incorporated into the liquid or liquefied binder system by planetary mixing accompanied by three roll milling. If a solvent-based product is required, the use of the solvent usually helps with the addition of the extender, although if the viscosity of the system is low some settling may occur.

4.4.4. Carbon Black

Carbon blacks can be produced by a number of different processes, each of which plays an important part in determining the quality of the material produced. Over 90% of the carbon black currently produced is made by the furnace process, in which oil or gas is thermally decomposed to form the carbon black particles. Carbon blacks produced by this process are usually almost neutral in pH. Other types of carbon black can be made with the differing processes, examples being thermal blacks, channel blacks, lampblack and acetylene black.

The main applications of carbon black are for ultra-violet light and thermal protection as well as the control of electrical conductivity. It finds considerable use in wire and cable sheathing, particularly with crosslinked polyethylene. Carbon black and its analogues are frequently also used as the black pigment for epoxide transfer moulding compounds at concentrations of 0·2% or less.

The principal uses of carbon black are:

(1) as a pigment
(2) to control electrical conductivity
(3) to enhance abrasion resistance
(4) to increase the tensile strength
(5) to increase modulus

4.4.5. Clay

Another extender occasionally used in epoxide formulations is clay. Clay is a hydrated alumino silicate mineral, also known as kaolin, which is available in a calcined, anhydrous form. The hydrated form is non-abrasive, chemically inert and has a large surface area. It is naturally acidic and this must be taken into account when formulating for applications where corrosion of metallic conductors may be encountered. Hydrated clay disperses readily in most formulations, particularly when dispersants or surfactants are used, and is utilised in epoxide transfer moulding compounds to allow control of flow.

The calcined clays are considerably harder than their hydrated counterparts but they impart better electrical properties to both thermosetting compounds and thermoplastics. Clay is also of particular use in fibre-filled epoxides, where its plate-like structure and viscosity increasing properties prevent fibres 'blooming' or appearing at the moulded surface. Calcined clay is used to increase volume resistivity in the formulation of electrical laminate type compounds. A hydrophobic surface treated version of calcined clay is available for applications where enhanced electrical properties are required even under humid conditions.

4.5. FLAME RETARDANTS

The extensive use of polymeric materials in all areas of the electronics industry, including consumer electronics, has reinstated the introduction of flammability regulations to safeguard the end user. Considerable concern has been expressed about the ability of synthetic polymers to burn and means have had to be found for controlling their flammability. This concern encompasses not only suppliers of plastic materials, but also the manufacturers of electronic components who have to ensure that the assemblies they sell to their customers do not represent a fire hazard. The requirements for reduced flammability often necessitate the addition of flame retardant additives into epoxide resin formulations and many materials are now available that provide valuable service in reducing their flammable nature.

In order to develop effective flame retardants, it has been necessary to carry out much work on the actual mechanisms occurring when a material burns and these will be briefly discussed before the currently available flame retardants are reviewed. When an object burns, several processes occur, including heating, degradation, volatilisation and oxidation. The flames produced during the burning process are sustained by the gas phase volatiles evolved from the decomposing material. Various methods are available for extinguishing the flames, such as cooling the solid, or altering the atmosphere available to the flame, so as to change the degradation and decomposition processes to give non-flammable products. Additives used in epoxide formulations function both in the solid phase and in the vapour phase. In the solid phase, flammability is reduced by altering the thermal degradation mechanisms or by forming a barrier at the surface, whilst in the vapour phase flammability is reduced by interfering with the oxidation processes.

The oxidation that occurs in the vapour phase is a free radical process and additives that can trap radicals such as $H^\cdot$, $O^\cdot$, $HO_2^\cdot$ and $OH^\cdot$ may be useful as flame retardants. Halogen-containing compounds are an example of particularly good radical scavengers and thus halogenated organic compounds find extensive use in flame retarded compound formulations.

4.5.1. Flame Retardant Epoxide Resins
Early efforts to impart flame retardancy upon epoxide resin formulations involved blending co-reactive halogenated compounds with liquid epoxies. These merely served as diluents and quite often resulted in products with markedly reduced physical properties such as T_g, thermal stability and compressive strengths. In order to avoid these problems, flame retardant epoxide resins have been synthesised containing halogen atoms within their molecular structures and these are now commonly used in epoxide resin based formulations. Their structures

are basically the same as the commonly used epoxide resins, except that some of the hydrogen atoms attached to the benzene rings have been replaced by halogen atoms such as chlorine and bromine. Typical examples are the halogenated derivatives of bisphenol A, for example

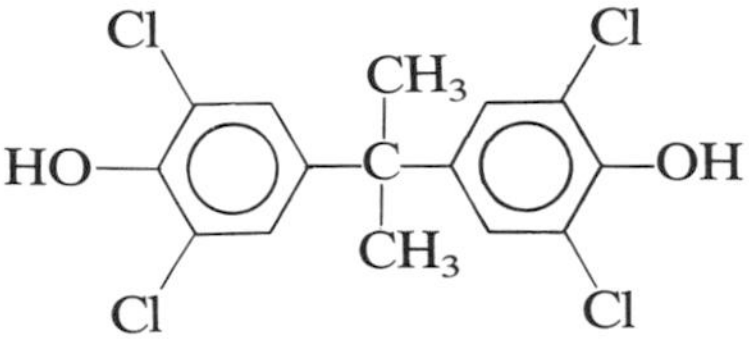

Tetrachlorobisphenol A 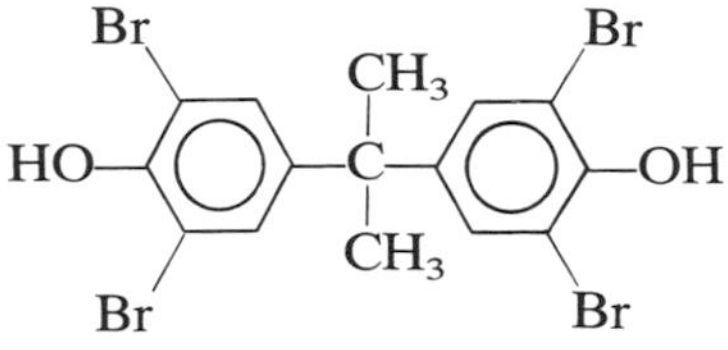   Tetrabromobisphenol A

When such halogenated epoxides decompose at high temperatures, as is the case during combustion, various halogen-containing species are formed, which help to trap the free radicals formed and their removal snuffs out the flame.

Although both brominated and chlorinated epoxide resins have been used, the preferred halogen is bromine. This is because less bromine (by weight) is required to impart flame retardancy than with chlorine. Another important reason for the selection of bromine in preference to chlorine is thermal stability.

Quantity of halogen required to impart flame retardancy upon a clear epoxide casting

Halogen	% (by weight)
Bromine	26–30
Chlorine	13–15

Quantity of halogen in tetra halogenated bisphenol A

Halogenated bisphenol A	% (by weight)
Tetrachlorobisphenol A	39
Tetrabromobisphenol A	59

Although the introduction of epoxide groups at either end of the bisphenol A molecule reduces the weight percentage of halogen present, it is still three times greater than is actually required in the case of the diglycidyl ether of tetrabromobisphenol A. This means that non-halogenated epoxide resins can also be used in the formulation giving some leeway in the final system composition. Apart from their use in structural laminate formulations, brominated epoxide resins find considerable use in both electrical and electronic applications as components of adhesive, coatings, encapsulant and potting compound formulations.

4.5.2. Antimony Oxides

For certain high reliability electronic applications it is desirable to have as little halogen as possible present in the epoxide resin formulation. One convenient way of reducing the halogen content of an epoxide resin system but maintaining the flame retardancy, is by the use of antimony oxides. The antimony oxide most commonly used is Sb_2O_3 although Sb_2O_4 and Sb_2O_5 also find occasional use. When an antimony oxide is used with halogenated resins a synergistic effect occurs, allowing a substantial reduction in the quantity of halogenated resin needed. Synergism is said to occur when the effects of two components combined is greater than the sum of their individual effects. On its own, antimony oxide has little effect as a flame retardant and may remain totally unaffected during the combustion of an epoxide. However, in the presence of an organic halide species the corresponding antimony halide is produced

$$6RX + Sb_2O_3 \rightarrow 2SbX_3 + 3R_2O$$
$$6HX + Sb_2O_3 \rightarrow 2SbX_3 + 3H_2O$$

Various hypotheses have been proposed in an attempt to explain how this synergistic flame retardancy occurs. One theory proposes that a reaction occurs between the polymer and the resultant antimony trihalide, which leads to char formation on the polymer surface. This char is formed in place of volatile, flammable gas and also reduces further volatile formation by acting as a thermal barrier to further polymer breakdown. The antimony halides formed during the combustion process are very volatile and they constitute a blanketing atmosphere over the epoxide, thus reducing the oxygen supply to the flame. Antimony halides are also very reactive, and a second theory proposes that, once they have volatilised into the flame, they decompose into antimony compounds and halogen free radicals. The antimony compounds function as energy dissipators and the halogen free radicals react with other free radicals altering the flame chemistry.

Irrespective of the mechanisms by which they work, antimony oxides allow considerable reductions to be made in the amount of halogenated resin required to impart a certain level of flame retardancy. A formulation ulilising an epoxide resin containing 14% bromine by weight would only need 5% of bromine to achieve the same flame retardancy if 3% of Sb_2O_3 was also added.

4.5.3. Alumina Trihydrate

Alumina trihydrate ($Al_2O_3 \cdot 3H_2O$) is another commonly used inorganic flame retardant, it is chemically inert and non-toxic. Polymer combustion is retarded by an exothermic dehydration reaction in which the hydrated alumina absorbs part of the heat of combustion and, being in the solid phase, acts as a heat sink removing energy from the combustion zone. The evolution of water vapour may also be important in diluting combustion gases and reducing the availability of gaseous oxygen. Alumina trihydrate is normally used in epoxide formulations between 40 and 60 weight % and it also imparts good arc and arc track resistance, which is useful in high voltage encapsulation work. Alumina trihydrate filled cycloaliphatic epoxide resin systems have been extensively used in transformer, insulator and switchgear applications. It is not generally thought to exhibit any synergistic effects with other flame retardants.

4.5.4. Other Flame Retardants

Whilst antimony oxide and alumina trihydrate are the two most commonly used inorganic flame retardants, a large number of other materials are available for use in speciality applications, or as cost-saving alternatives. Concentrations of borates, in excess of 20% by weight, can impart self-extinguishing properties upon certain epoxide resin based formulations. Zinc borate is most commonly used, although sodium, calcium, ammonium and lead borates also find considerable use. Organic boron compounds such as trimethoxyboroxine, borolene and borinane are particularly useful because of their solubility in the resin system.

Salts such as ammonium polyphosphate have been successfully utilised as flame retardants in filled epoxide resin systems. It is more usual to employ phosphorus/halogen mixtures in epoxide resins. A combination of 1·5–2% phosphorus with 5–6% of chlorine will impart self-extinguishing properties to epoxides.

Finally, apart from using halogenated epoxide resins or additives, it is also possible to use halogenated curing agents, examples being

chlorendic anhydride, tetrabromophthalic anhydride and dichloro-
maleic anhydride

| Chlorendic anhydride | Tetrabromophthalic anhydride | Dichloromaleic anhydride |

When considering the use of chlorinated flame retardants, considerable thought should be given to the reliability performance required of
the encapsulated or moulded part, since chloride ions generated from
these materials may have deleterious effects on certain metallic conductors. It should also be remembered that the halogenated anhydride
curing agents will tend to be more reactive than their non-halogenated
counterparts owing to the electron withdrawing effects of the halogen
atoms.

4.6. OTHER USEFUL CONSTITUENTS

Even though there are literally hundreds of different epoxide resins
and hardeners commercially available, giving countless formulations
with a diversity of cured properties, further modifications of a material
are sometimes still required. These modifications can be achieved by
incorporating a range of compounds known as diluents, flexibilisers
and plasticisers into the epoxide resin system during the formulation
process. The properties of some typical diluents, flexibilisers and
plasticisers are now discussed.

4.6.1. Diluents
For many uses of epoxide resin systems, a low viscosity product is
required and the role of the diluent is usually to reduce the viscosity of
the formulated compound. Diluents can also be used to tailor other
characteristics of the epoxide, such as the chemical resistance, electrical properties, thermal expansion coefficient and glass transition temperature. They may be conveniently divided into two groups, the
reactive diluents and the non-reactive diluents.

Examples of non-reactive diluents are the aromatic hydrocarbons such as xylene and toluene

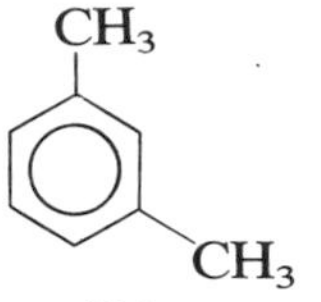

m-Xylene    Toluene

These are used in concentrations of less than 5 phr and have little effect upon the cured physical properties, but there may be more significant changes in the crosslinked products' chemical and solvent resistance. Obviously, as the concentration of non-reactive diluent is increased, the effect upon the cured properties becomes more pronounced. Care also has to be taken with thermally cured systems that the curing temperature does not cause volatilisation of the diluent, as this can result in the formation of voids and blisters in the cured material.

One of the most popular groups of all the non-reactive diluents is comprised of the phthalate esters, which have found extensive use in many types of other polymer formulation as well as with epoxides. The properties of several commonly used esters are given in Table 4.6.

The most widely used phthalate ester is probably dibutyl phthalate, which has the following structure:

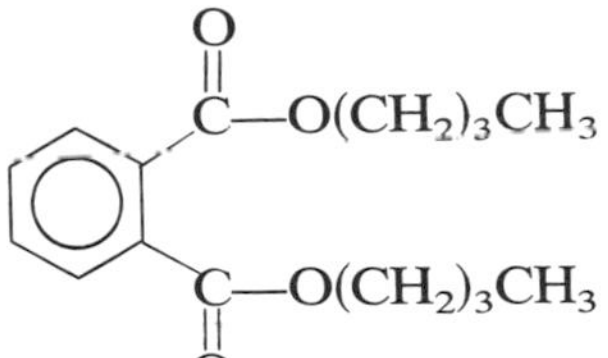

Dibutyl phthalate

Table 4.6
Properties of some Commonly used Phthalate Esters

Ester	Formula	Molecular weight	Boiling point (°C)	Density (g/cm³)
Diethyl phthalate	$C_6H_4(COOC_2H_5)_2$	222	249	1·12
Dibutyl phthalate	$C_6H_4(COOC_4H_9)_2$	278	310	1·05
Dihexyl phthalate	$C_6H_4(COOC_6H_{13})_2$	334	345	1·01
Dioctyl phthalate	$C_6H_4(COOC_8H_{17})_2$	390	340	0·98

When used in the range of 15–20 phr with DGEBA-based resins, four-fold reductions in viscosity can be achieved, although, by nature of its chemical structure, it may also bring about a reduction in crosslinking density. Other materials that are useful under certain circumstances as non-reactive diluents are the low molecular weight styrenes and the bisphenol- and phenolic-based resins. Before selecting a non-reactive diluent for use in an epoxide resin formulation, careful consideration should be given to the desired final properties, since although certain properties may be improved others may be deleteriously affected.

4.6.2. Reactive Diluents

Reactive diluents are usually small molecules that contain a single epoxide group in their structure, enabling them to react during the curing reaction. The fact that they are monofunctional epoxides means that they reduce the crosslinking density of the cured products and this is illustrated by the resulting lower glass transition temperatures and reduced thermal stability. Examples of commonly used epoxide containing reactive diluents are shown below.

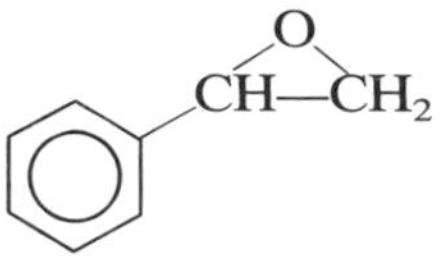

Styrene oxide

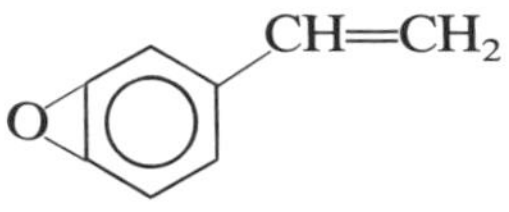

Vinyl cyclohex-3-ene oxide

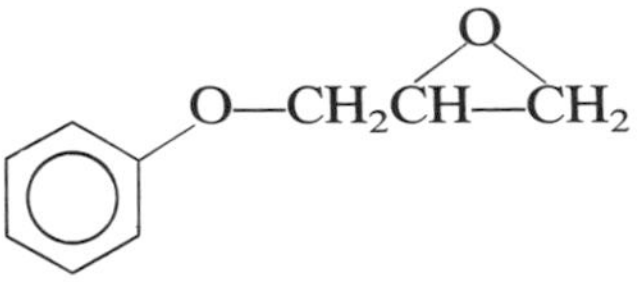

Phenyl glycidyl ether

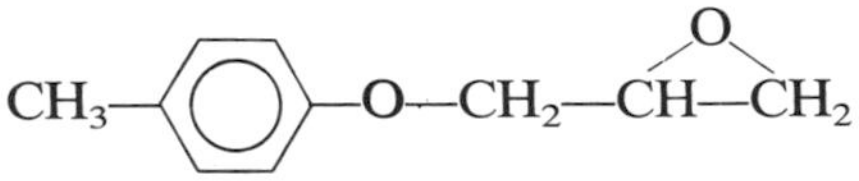

Cresyl glycidyl ether

A further group of monofunctional reactive diluents is the glycidyl ethers of C_{12}–C_{14} linear aliphatic alcohols. These materials have a higher molecular weight than butyl glycidyl ether, which has been used extensively in the past. This higher molecular weight results in odourless compounds with low volatility in air and reduced potential for skin irritation. In general, however, reactive diluents should be handled with care, since they are sometimes more prone to inducing dermatitis than the epoxide resins themselves.

4.6.3. Flexibilisers and Plasticisers

For some applications there may be a requirement for a normally brittle and hard epoxide system to be modified to have greater flexibility and toughness. There are essentially two ways in which this can be achieved and a typical example would be to convert a rigid system into one that has improved thermal shock and impact resistance for castings, improved flexibility for films or enhanced peel strength for adhesive applications. The first method of achieving this goal is to incorporate molecules into the formulation that have long flexible chains and which can be reacted into the polymer network during curing. This method is known as flexibilisation and flexibilised resins, curing agents or additives can be used. The second method is to again incorporate long chain flexible molecules into the formulation, but this time no reaction takes place and the molecule is not chemically linked into the crosslinked cured polymer. This method is known as plasticisation.

The flexibilising epoxide resins may be either monofunctional or polyfunctional in a similar manner to the reactive diluents. The monofunctional epoxide resins are usually derived from molecules containing long chains of atoms such as the vegetable oils and other molecules containing aliphatic chains with more than 12 carbon atoms. These compounds are epoxidised and, during the curing reaction, their epoxide groups react with the growing polymer network effectively inhibiting further propagation of the polymer chains. This leads to the formation of long floating chains within the crosslinked matrix, which serve to increase the volume of the crosslinked product, allowing greater movement of the material under imposed loads. To a large extent, the monofunctional diluents can also be considered as flexibilisers.

The polyfunctional flexibilisers are comprised largely of difunctional epoxides derived from molecules with flexible structures. The commercially available flexibilising resins are commonly based upon the glycidyl derivatives of glycols and dimerised acids. The general structure of an epoxidised glycol is

$$CH_2-CH-CH_2(O-CH_2-\overset{\overset{\displaystyle R}{|}}{CH})_n-OCH_2CH-CH_2$$

The value of n would be in the range 2–7.

The diglycidyl derivatives of fatty acids are also good flexibilising resins and such materials generally have viscosities of between 400 and 600 cP at room temperature. They can be used with most epoxide curing agents, but their incorporation into a system generally leads to long pot lives and the need for longer cure times.

Flexibilisation of a formulation can also be achieved by using flexibilising curing agents or other additives. These materials are similar in their basic structures to the flexible epoxide resins and usually have reactive end groups separated by flexible long chain molecular segments. A typical example is dodecenyl succinic anhydride which has the structure

$$Me(CH_2)_2—CHMe—CH_2—CMe{=}CH—CMe_2—CH—CO$$
$$CH_2—CO \diagdown O$$

In certain applications this material can impart outstanding electrical properties along with increased flexibility. When it is used in a 50/50 adduct with hexahydrophthalic anhydride it gives a useful eutectic mixture with a melting point of $\sim 0°C$. One other group of flexibilising hardeners also worthy of mention is the polysulphides. They have a polymeric structure which is terminated at each end with a thiol group (–SH). Although the thiol group can react slowly with the epoxide resins, they are generally used in conjunction with catalysts such as DMP10 and DMP30. The polysulphides have found considerable usage in the formulation of epoxide systems for the encapsulation and potting of delicate components.

4.7. GENERAL FORMULATION CONSIDERATIONS

Having been presented with the large number of materials that are commonly used in the formulation of epoxide encapsulants, the reader new to this area may be bewildered at the vast range of possibilities for the composition of a given compound and may be wondering how one begins to select the individual components to impart the desired properties in the cured product. It is relatively easy to define the properties required of an epoxide for a specific application, but it is more difficult to select ingredients that will allow all of these desired properties to be accommodated and, quite often, some degree of compromise is necessary. The formulator will have to make a choice of components from a range of literally hundreds of available resins, hardeners, catalysts and other ingredients, each of which is often available in a variety of forms, in an attempt to balance their contrasting properties. The properties of a given formulation can be conveniently divided into those required before, during and after cure.

The main characteristics of the material required before cure that will influence ingredient selection are:

(1) The overall cost limitations for the product.
(2) The required physical form of the compound. Is a liquid or a solid required?
(3) Chemical stability. What storage life is required and does the compound have to be one or two parts?

Having taken these characteristics of the compound into account, the properties of the compound during the curing reaction now have to be considered. The major factors here are:

(1) What cure schedule is possible? Does the compound have to cure within a specific time? Is there a maximum permissible curing temperature?
(2) Are the components of the formulation chemically compatible with the substrate being encapsulated and coated?
(3) How much cure shrinkage can be tolerated?
(4) What viscosity is required for both liquid systems and moulding compounds?
(5) What curing exotherm is tolerable under the determined curing conditions?

Finally, we come to the properties of the cured product. Many of these properties will obviously be influenced by or influence the properties required before and during curing of the compound. The more important characteristics commonly considered when formulating epoxides for electronic applications are:

(1) Colour and optical properties; an epoxide may need to be a specified colour or it may need good transmissive properties at a given wavelength for electro-optic applications.
(2) Density; there may be limitations on the weight of a product, particularly if it is intended for aerospace use.
(3) Thermal expansion coefficient (α_1); the thermal expansion coefficient mismatch between the epoxide and a substrate may need to be minimised.

(4) Adhesion; good adhesive properties may be required, as in die attach applications.

(5) Thermal conductivity; high thermal conductivity may be required for heat dissipation from power devices or very highly integrated circuits.

(6) Chemical and solvent resistance; good chemical and solvent resistance may be required for epoxides exposed to harsh chemical environments.

(7) High purity; this is required for epoxides in contact with delicate microcircuits where impurities may cause corrosion of conductors to occur.

(8) Low gas, moisture and vapour permeability; again, this is required for epoxides in contact with delicate microcircuits, where corrosion or parametric shifts may occur.

(9) Internal stress; low internal stress may be important with epoxides in contact with large microcircuits, microcircuits with multi-layer metallisations and large pin count packages.

(10) Low alpha particle emission; ingredients may need to be pure with respect to actinide elements, to avoid alpha particle emissions that can cause soft errors in dynamic random access memories of 64K capacity and upwards.

(11) Thermal stability; stability at high temperatures may be required for an epoxide operating in a harsh environment where elevated temperatures are experienced or with high power devices.

(12) Tensile and flexural strength, hardness and glass transition temperatures (T_g) are all important properties where good mechanical performance and physical integrity is required.

(13) Flammability; an epoxide may need to be flame retarded to meet certain customer requirements. Examples would be to achieve Underwriters Laboratory listings (UL), such as UL94 $\frac{1}{8}$ in V.O.

(14) Electrical properties such as volume resistivity, dielectric constant, dielectric strength, arc resistance, dissipation factor and loss factor are all important where the epoxide is in contact with high voltages or where there is intimate contact with a microcircuit surface. Certain of these electrical properties are particularly important with high-speed devices operating in the frequency range above 10 MHz, where low values are generally desirable.

Only by carefully considering all of the properties that are required of a compound formulation before, during and after cure will it be possible to make a final selection of the ingredients required to give a material suitable for a specific application.

4.8. SUMMARY

This chapter has attempted to describe the principal ingredients more commonly used in epoxide formulations and to illustrate the vast selection of choices available to the prospective compound formulator. The important factors that may need to be taken into account for a given application have also been reviewed. For further information and assistance with formulation, the reader is referred to the list of further reading at the end of this chapter.

REFERENCES

1. Farben, I. G., German Patent 676,117 (1939).
2. Burhans, A. S., *Insulation/Circuits* (Jan. 1979) 37.

FURTHER READING

Abshier, C. S., Berry, J. and Maget, H. J. R., Toughening agents improve epoxy encapsulants. *Insulation/Circuits*, **27** (Oct. 1977).
Anon. Flame retardants. *European Plastics News*, **10** (Nov. 1982).
Bruins, P. F. (Ed.), *Epoxy Resin Technology*. 1968, John Wiley and Sons Inc., New York.
Buttrey, D. N., *Plasticisers*. 1960, Franklin Publishing Company, Inc., Palisade (New Jersey).
Hirsch, H., Resin systems used for encapsulation of microelectronic packages. *Proc. Soc. Plastics Engineers, 28th Annual Technical Conference*, 380 (May 1970), New York.
Katz, H. S. and Milewski, J. V., *Handbook of Fillers and Reinforcements for Plastics*. 1978, Van Nostrand Reinhold Company, New York.
Keirsey, R. A., Evolution of epoxy resins in semiconductor device encapsulation. Ann. Pack. Tech. Conf. Disp. (Tech. Pap), Society of Plast. Eng. 1981, *6th Reinforced Plast. Comp. Adhes. Thermosets*, pp. 172–9.
Lee, H. and Neville, K., *Handbook of Epoxy Resins*. 1967, reissue 1982, McGraw-Hill, New York.
Lyons, J. W., *The Chemistry and Uses of Fire Retardants*. 1970, John Wiley and Sons Inc., New York.

Markstein, H. W., The use of plastics and resins in electronics. *Electronic Packaging and Production,* **51** (Aug. 1981).

Nielson, P. O., Properties of epoxy resins, hardeners, and modifiers. *Adhesives Age,* **42** (April 1982).

Oldberg, R. C., The effects of epoxy encapsulant composition on semiconductor device stability. *J. Electrochem Soc: Solid State Science,* **118,** 129 (Jan. 1971).

Potter, W. G., *Epoxide Resins.* 1970, Butterworth and Co. (Publishers) Ltd., London.

Reinhart, J., Epoxy moulding compounds for semiconductor encapsulation. *Circuits Manufacturing,* **29** (Aug. 1977).

Schultz, N., Potting and encapsulating electronic components. *Adhesives Age,* **19** (July 1983).

Snyder, P. M., Developments in semiconductor device encapsulation, *Microelectronic Manufacturing and Testing,* **41** (Sept. 1983).

Suzuki, H., Sato, M., Muroi, T. and Watanabe, Y., *New Moulding Compounds for Electronic Devices.* Soc. Plast. Eng. Tech. Paper 19, 6 (1973).

Chapter 5

PLASTIC ENCAPSULATION OF SEMICONDUCTORS BY TRANSFER MOULDING

MARTIN T. GOOSEY

Dynachem Corporation, Tustin, California, USA

5.1. INTRODUCTION

In the early days of semiconductor fabrication, devices were largely packaged in glass, ceramic or metal. However, by the 1960s device manufacturers began to seriously investigate the possibility of encapsulating their products in some of the new plastic materials that were becoming available, as an alternative to the more conventional methods. The main incentives for this move towards plastic encapsulation were the large cost reductions offered over the standard ceramic and metal parts and compatability with mass production techniques, which oftcn allowed several hundred parts to be encapsulated in a single operation taking less than 2 min. Motorola introduced the plastic T092 package in the mid 1960s and by the late 1970s worldwide annual production of plastic encapsulated parts was estimated to have exceeded 10 billion units. The total market in the USA for encapsulants of both active and passive components was valued at well over US $80 million in 1980 and is predicted to have exceeded US $225 million by 1985.

There are several ways in which a semiconductor device can be encapsulated to give it good physical integrity and environmental protection. The three most commonly used techniques are transfer moulding, potting and conformal coating, with transfer moulding being the most popular method. In this process, the components to be encapsulated are placed into the cavities of a mould and a moulding

Martin T. Goosey

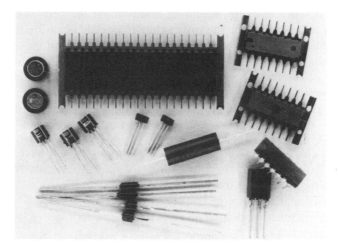

Fig. 5.1. A selection of commonly transfer moulded devices.

compound, liquefied by heat and pressure, is forced into the cavities where it solidifies giving the familiar plastic encapsulated device. Typical plastic encapsulated parts are shown in Fig. 5.1.

Whilst the materials and machinery used in the early 1960s can be considered fairly crude, considerable progress has been made in both moulding equipment design and compound formulation. The main areas where improvements were required were in device reliability, production rates and material wastage. The major achievements of the equipment manufacturers in both mould and press design have allowed the number of parts moulded per mould cycle to increase dramatically, so that in some cases 400–500 parts can be finished per shot. Much larger active devices with higher pin count packages can now be moulded, while the amount of material wasted in the runner systems has been reduced. Most recently, several manufacturers have begun offering automatic moulding systems, which remove the manual labour content from the moulding operation, allowing mould cycle times of less than a minute to be easily achieved.

By far the most commonly used materials for transfer moulding both active and passive components are those based upon epoxide resin chemistry. In the past, epoxides were generally in competition with other thermosetting materials, such as the phenolics and diallyl phthalates, in the moulding of structural parts, bushings and cases. However, the longer flowing epoxide moulding compounds for semiconductors are now largely unrivalled, the only other materials occasionally used in these applications being silicones, silicone–epoxides and more rarely polyurethanes, polyamides and polyphenylene sulphide. (In 1980 about 90% of the total USA consumption of semiconductor encapsulants was epoxide resin based.) The main reason for the popularity and increasing use of epoxides in this field, at the expense of other materials, is that they possess several distinct advantages such as:

(1) Excellent compatibility with other components of the formulation, for example good bonding to filler, particularly when coupling agents are used.
(2) Dimensional stability in the cured product.
(3) Good chemical and solvent resistances due to high crosslinking density.
(4) Retention of excellent electrical and mechanical properties at elevated temperatures, in some cases as high as 180°C.
(5) Cure by an addition reaction, avoiding the evolution of volatile condensation molecules.
(6) Relatively low shrinkage.
(7) Good hot hardness.
(8) Mouldability at lower than conventional transfer pressures, avoiding wire bond sweep on delicate integrated circuits.
(9) Rapid curing, allowing the high-speed production of large numbers of parts.

However, as with most materials, there are also some disadvantages, the major two being:

(1) High adhesive strength; epoxides are good adhesives and release agents have to be employed to prevent parts from sticking in a mould.
(2) Epoxides are inherently flammable and require the incorporation of flame retardant additives if flammability control is required.

Just as the moulding equipment suppliers have improved their products, so have the moulding compound manufacturers and distinct improvements have been made over the last few years in epoxide moulding compound formulation. Epoxide moulding compounds now exist with properties that were totally unattainable only a few years ago, allowing their use in many new areas, for example:

(1) Low stress materials are available that enable the moulding of large pin count devices and large package sizes without the familiar problems caused by stress such as cracking of passivation layers, moved metal phenomena and package warping.

(2) Low alpha particle emission materials are available enabling the encapsulation of 64 and 256K dynamic RAMs without the fear of soft error generation due to alpha particle emissions from the moulding compound.

(3) Faster curing materials are available that are compatible with the latest generation of automated moulding machinery, allowing mould cycle times of around 45 s to be easily accommodated.

(4) Laser markable materials are available that allow identification numbers to be burnt onto the surface of a package, removing the need for an ink marking process.

In addition to these developments for specific applications, many other general improvements have been introduced in the last few years, the principal examples being:

(1) Enhanced shelf-life during storage.
(2) Higher as-moulded and post-cured glass transition temperatures.
(3) Reduced mould staining due to the introduction of higher purity and synthetic release systems.
(4) Reduced flashing and bleeding.
(5) Reductions in impurity levels in all raw materials used.
(6) Improved hot hardness.

As a consequence of these and continuing advances being made in moulding compounds, the packaging of components by transfer moulding is increasing in popularity and will remain by far the most important packaging technique in terms of volume for the foreseeable future.

5.2. TRANSFER MOULDING EQUIPMENT CONSIDERATIONS

Transfer moulding was the term chosen by L. E. Shaw, the originator of the process, 'to describe the method and apparatus for moulding thermosetting materials, whereby the material is subjected to heat and pressure and then forced into a closed mould cavity by this same pressure and held there under additional heat and pressure until cure is complete'. Transfer moulding equipment for electronic components consists essentially of a hydraulic press equipped with platens, one of which contains a chamber known as a pot, in which the moulding powder is placed and liquefied by a combination of pressure and heat. A piston or ram transfers the molten material into cavities of the mould via a series of channels known as runners. A typical transfer moulding press is shown in Fig. 5.2.

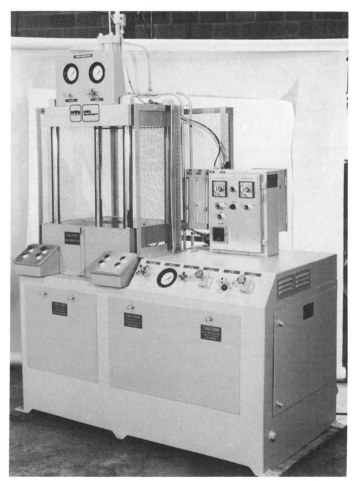

Fig. 5.2. Transfer moulding presses are available in many sizes and capacities. Illustrated is a 75 ton model sold by MTI Equipment, Ivyland, PA.

The moulding compound can be introduced into the pot as a ground powder or more usually as a preheated preform. In this case the powder is compressed to form a solid pellet of material which is dielectrically heated to around 90°C using a high-frequency preheater. The preheated preforms are loaded into the pot and transferred by the ram at pressures typically around 1000 psi, although as little as 50 psi may be sufficient for some specially formulated epoxides. For an epoxide compound used for moulding many tens of semiconductor devices in a single shot, a transfer time of between 15 and 20 s is usually typical, although for some compounds and moulds, operation outside these limits may be necessary. The mould temperature would normally be around 175°C and the parts could be removed after a mould cycle time of $1\frac{1}{2}$ to 2 min. The mould is usually designed so that, when it opens, ejector pins in the cavities and runners push the moulded devices free. The moulded strips of components will then be removed from the mould and subjected to a number of finishing processes such as deflashing, trimming and forming of the lead-frames, post-curing and marking before the final usable product is obtained.

Transfer moulding can be successfully used for the packaging of resistors, resistor networks, capacitors, diodes, transistors, integrated circuits, coils, reed relays and many other discrete components. The process only becomes economically feasible, however, for mass production operations since the initial high cost of the press, moulds, preheater and preformer can only be amortised over the production of many hundreds of thousands of parts.

In order to successfully design a mould for the transfer moulding of semiconductors, it is essential to define the basic parameters and processing variables that are likely to be important in order to gain a thorough understanding of just what is actually required before the mould can be manufactured. A mould will generally be developed specifically for the device to be moulded and the materials to be used. Some of the important basic factors which need to be taken into account before a mould design can be initiated are:

(1) The shrinkage of the moulding compound during cure.
(2) The likely mould release behaviour of the moulding compound and what draft angles will be needed to enhance component release.
(3) The siting of gates to achieve optimised filling of cavities and to give good cosmetic appearance.

(4) Whether or not ejector pins will be required to release the moulded parts from the mould and, if so, where they must be located, taking into account the rigidity and hot hardness of the cured epoxide and the sensitivity of the encapsulated devices to stress.

(5) Degree of venting required to prevent air and volatile entrapment in the moulding.

(6) Whether the surface of the moulded components needs to be glossy or matt and whether conventional marking or laser marking is to be used.

5.2.1. Gate Location

The positioning of the gate or gates that admit the molten moulding compound into the cavity of the device to be moulded is determined by a number of considerations. A gate must permit proper flow of the epoxide as it enters the cavity and it must also be situated so that the moulded parts can easily be removed from the mould at the completion of the moulding cycle. It should also be situated to facilitate easy cleaning as and when necessary. For the transfer moulding of dual in-line packages a single gate is usually sited at the narrow end of the package and is offset to one side in the parting line area. The depth of a gate in a semiconductor application would typically be in the range $0.010-0.015$ in.

5.2.2. Venting

All transfer moulds have to be vented in order that any air or volatiles in the mould cavities can escape, preventing the build up of back pressure and allowing complete void-free filling. The size and location of the vents will depend upon the size and design of the moulding and the position of any pins or inserts. A vent is simply a small groove machined from one face of the mould that allows the expulsion of unwanted gaseous products from the cavity, but is too narrow to allow the passage of any of the moulding compound. For the transfer moulding of dual in-line packages a single vent is usually machined at the opposite end of the package to the gate, as this is the point which fills last with compound. Typical dimensions for a vent of this type would be between 0.002 and 0.005 in deep and $\frac{1}{4}-\frac{1}{8}$ in wide, although it is normal practice to begin with a minimum gate dimension and then enlarge it as necessary.

5.2.3. Runner Systems

The runner system of a transfer mould refers to the series of channels that allow the distribution of the moulding compound to the separate chases and eventually the individual cavities. The runners emanating from the plunger/cull area which feed each individual chase of the mould are known as the primary runners and those that distribute the material from the primary runners to each individual cavity are known as secondary runners. A typical primary runner would be hemispherical or trapezoidal in shape with a width and a depth of 3–5 mm. The secondary runners are generally around 25–35% smaller in area than the primary runners. Runners are usually machined into the ejection side of the mould and are often equipped with their own ejector pins. Whenever possible, the runners are kept short to enhance the material flow and reduce the transfer time required to fill the cavities.

5.2.4. Parting Line

The surface at which the two halves of a mould separate is known as the parting line. The optimum parting surface, as far as mould construction is concerned, is a flat plane which remains flat after all moulded parts and projections are removed. This is because such a flat surface can then be easily machined flat again, should any wear or damage occur. For a standard chase mould for moulding dual in-line integrated circuits, the parting line is at the centre of the package in the plane of the lead-frame.

5.2.5. Shrinkage and Draft Angles

All thermosetting moulding compounds shrink to a certain extent when they cure and thus allowances must be made for this shrinkage during mould construction. The mould cavities are made oversized by an amount equal to the shrinkage of the cured compound. Typical shrinkage allowances range from 0·004 to 0·010 m/m and obviously the use of a material with different shrinkage characteristics to the one the mould was made for may lead to dimensional tolerance problems.

Draft angles are the degree of taper machined into a cavity wall in order to facilitate easier moulded-part removal. By the use of tapers, the distance required to totally free the moulded part from the mould is significantly reduced from the case where the walls of the cavity are vertical. Less force is required from ejector pins in order to free the moulded device and less warping or flexing of delicate mouldings is likely to occur giving increased yields for particularly sensitive devices.

The area of material shrinkage and draft angles has become more important in the last few years because the increasing size of both chips and packages has meant that many devices are susceptible to internal stresses. These stresses are primarily caused by thermal expansion coefficient mismatches occurring between the moulding compound and the lead-frame and device. One way of reducing these stresses is to modify the epoxide moulding compound formulation to reduce its thermal expansion coefficient and shrinkage. The normal draft angle required for standard moulding compounds is typically around 4°, whilst in order to obtain good release from the mould with the latest generation of low stress compounds draft angles of 8–12° are now recommended.

5.2.6. Mould Finish

The finish of the cavity surface of a mould is particularly important because it determines the quality of the finish of the moulded part. Moulds are quite commonly hard chrome plated to seal any porosity in the steel mould surfaces. Since epoxide moulding compounds are mineral filled, they tend to be extremely abrasive and so hard chrome plating can give an effective decrease in the rate of wear of the runners

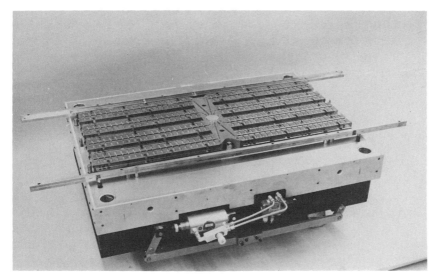

Fig. 5.3. A 320 cavity MTI encapsulation mould for 14/16 pin dual in-line ICs.

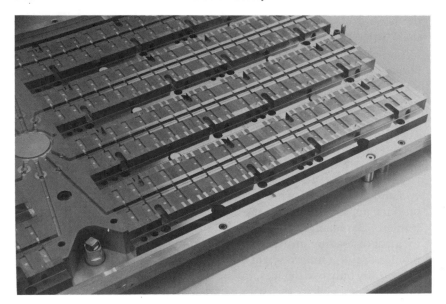

Fig. 5.4. A close-up view of the 320 cavity mould illustrating the position of gates, knock-out pins, runners and runner turn offs.

and cavities. It is essential that a surface to be hard chrome plated is very highly polished because the plating will effectively duplicate the topography of the underlying steel. Hard chrome plating also tends to prevent corrosion of the mould and allows a smoother flow of the material into the cavities.

Having taken all of the above factors into account, the mould manufacturer can then produce a workable design for a mould. Figures 5.3 and 5.4 show the layout of a typical 320 cavity mould for 14/16 pin dual in-line ICs.

5.2.7. Automated Moulding Systems

A number of manufacturers, principally in Japan, are now offering automated moulding systems in an attempt to reduce labour costs and to increase the productivity of semiconductor device moulding operations. In the past, many major semiconductor manufacturers were forced to abandon their indigenous assembly operations because of high labour costs. As a consequence, much of this so-called back end, or final, assembly work has, until recently, been performed at low-cost assembly houses in the Far East which rely heavily upon cheap labour to give assembly costs which could not be achieved in either Europe or the USA. In the last few years, labour costs have increased much faster than expected at these plants and, as a consequence, many manufacturers now feel that final assembly and moulding can be achieved just as economically onshore, with the added bonus of a faster turn around time than was attainable with shipment offshore. In order to maximise the productivity of in-house assembly, new assembly facilities are being equipped with the latest generation of automated moulding machines. A typical example is Motorola, which has automated moulding machines operating at its USA1 plant in Chandler, Arizona, and which it is planning to duplicate in East Kilbride, Scotland.

The Japanese automated moulding machines are totally enclosed systems that are fed with assembled leadframes at one side, the final product appearing at the opposite side ready moulded, trimmed and formed. The leadframes are placed in the mould cavities automatically by a robot loading arm which is fed from a cassette or magazine. The moulds are of a multiplunger design, with each plunger having runners to typically only two devices. Each plunger is automatically fed with mini preforms or tablets or, with some equipment, a specially ground powder. After loading the preforms and transfer moulding, the devices are automatically removed from the mould and transferred by a conveyor system to an automated trim and form operation, which is part of the equipment, or to storage cassettes ready for trimming and forming on a separate machine. After each mould cycle a cleaning operation is performed by revolving brushes.

The principal advantage of these automated presses, apart from the fact that they can be largely unsupervised when operational, is the increase in throughput that is achievable. Conventional moulding is limited not only by the cure time of the moulding compound but also by the speed at which the operator can load and unload the machine. The main automated moulding machine manufacturers are Towa, Yamada and Daichi Seiko of Japan and Fico of Holland (who are reported to have adopted a totally different approach to automated moulding by using moving moulds). A number of American manufacturers are also actively pursuing the manufacture of automated moulding systems.

5.3. TRANSFER MOULDING COMPOUND FORMULATION CONSIDERATIONS

The development and production of a typical epoxide moulding compound has usually been determined, as with other epoxide formula-

Table 5.1
Typical Components of an Epoxide Moulding Compound

Major components	*Minor components*
Epoxide resins	Release agents
Hardener resin	Pigments
Catalyst or accelerator	Flame retardants
Fillers	Additives

tions, by the properties desired for the compound, the availability of raw materials to manufacture the compound and the compound's ultimate selling price. From a practical point of view, the final product is often the result of a series of trade-offs or compromises between the various properties of the ingredients used in the formulation of the compound.

A typical epoxide moulding compound contains a variety of major and minor components, and the general ingredients found in most products are shown in Table 5.1. Each of these components is available with a variety of both physical and chemical properties, all of which have a distinct role in determining the final properties of the formulated product. The major influences these various constituents have on a compounds properties are now discussed in more detail.

5.3.1. Filler Content and Resin Viscosity Effects on Spiral Flow
The spiral flow of a moulding compound is a property that determines whether or not a mould will fill under given moulding conditions and whether the compound will have acceptable flash and bleed characteristics. The spiral is moulded under a given set of moulding parameters with a charge of material sufficient to give a cull within a specific thickness range. Moulding compounds for semiconductors will usually have a spiral flow in the range of 20–40 in. The spiral flow of a typical epoxide moulding compound varies with the viscosity of the resins and the amount of filler used in the formulation, as can be seen in Fig. 5.5.

The resin viscosity shown on the left-hand axis in Fig. 5.5 is an arbitrary measure of viscosity and may be taken to represent the predominant resinous component or it can be the net viscosity of the combined resin system. In either case, as the viscosity increases the spiral flow decreases. This effectively means that the viscosity range of the resins that may be employed is limited, to a large extent, by the

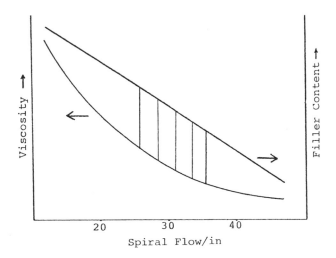

Fig. 5.5. Variation of spiral flow with viscosity and filler content.

desired spiral flow for the finished products (shown here as 25–35 in). The choice of resin viscosity is, however, also related to other factors besides spiral flow and these impose a further limitation upon the range of usable viscosities. In particular, the use of very high viscosity resins can cause wire sweep problems during moulding, whilst low viscosity resins can generate a resin bleed problem.

The right-hand vertical axis in Fig. 5.5 represents filler content and it can be seen that spiral flow decreases with increasing filler content. The spiral flow requirement thus has a direct bearing upon the filler content employed in most compounds, giving a certain range of acceptable filler concentrations. Another factor which is influenced by the filler content of a compound is the cure shrinkage with high filler contents giving lower values. Other aspects of the filler composition can also add further limitations to the development of a compound. For example, the filler particle size and particle size distribution affect the flash and resin bleed characteristics of the finished compound. Oversized filler particles can cause gate blockage resulting in incomplete filling of cavities. This is particularly important when moulding small-sized packages such as chip carriers and SOT23 devices, where gate depths may be thinner than for conventional transistor and dual in-line packaged integrated circuit moulds.

5.3.2. Thermal Expansion Coefficient and Filler Content

The thermal expansion coefficient of silicon is ~3 ppm/°C, whilst that of an unfilled epoxide is typically 50–60 ppm/°C. This difference results in a thermal expansion coefficient mismatch between a moulding compound and a semiconductor die (and also the leadframe material). Thermal expansion coefficient mismatch can lead to the build up of stresses in a device, which results in premature failure by cracking of passivation layers, failure of wire bonds and even cracking of the chip or package. One of the main uses of fillers is to reduce the thermal expansion coefficient mismatch between the epoxide and the encapsulated parts. For a given epoxide, the thermal expansion coefficient below the glass transition temperature (α_1) decreases with increasing filler content, as can be seen in Fig. 5.6.

Of greater significance, however, is the selection of filler and its effect upon the thermal expansion coefficient. For a specified loading, much lower values are achieved with fused silica than with α quartz. The obvious choice of filler to give a low thermal expansion coefficient at a given loading would be fused silica; however, there is a trade-off in properties, since the filler giving the best thermal expansion characteristics does not give the best thermal conductivity.

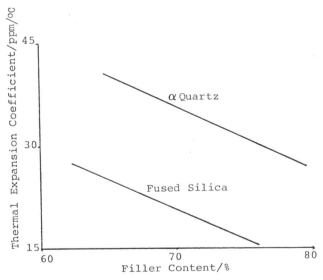

Fig. 5.6. Variation of thermal expansion coefficient with filler content.

5.3.3. Thermal Conductivity (λ) and Filler Content

Unfilled epoxides have relatively low thermal conductivities compared to the inorganic fillers and thus the incorporation of a filler such as silica into a formulation not only reduces the thermal expansion coefficient but also increases thermal conductivity. The variation of thermal conductivity with filler content for a typical moulding compound is shown in Fig. 5.7.

However, the most significant factor is again the choice of filler with α quartz giving higher thermal conductivity than is achieved with fused silica at the same loading. While the choice of filler for highest thermal conductivity is obviously in favour of α quartz, by considering both Figs 5.6 and 5.7 the trade-off between thermal expansion coefficient and thermal conductivity can be appreciated. In effect, for the highest thermal conductivity α quartz would be the best filler, whereas for the lowest thermal expansion coefficient fused silica would be selected. The choice of filler will therefore have to be carefully considered for a particular application depending upon which property is considered most important. It is of course possible to blend the two fillers to give a compromise in properties where moderate expansion and thermal conductivity properties are required.

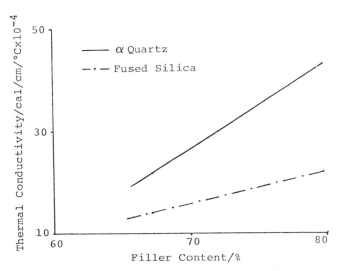

Fig. 5.7. Variation of thermal conductivity with filler content.

5.3.4. Mould Release Agents: Staining and Releasability

Epoxides are generally good adhesives (a property imparted by their molecular structure) and epoxide moulding compounds are no exception. Consequently, in order that moulded parts can be removed from a mould cavity, a mould release agent is usually included in the formulation. One of the commonest release agents is carnauba wax, a natural product, which unfortunately contains a small amount of impurities that eventually lead to staining of the mould and degraded cosmetic appearance of the moulded parts. There are thus conflicting requirements with the use of this wax: enough must be added to release any sticking tendency, yet the smallest amount possible should be used to avoid early staining. Figure 5.8 illustrates the relationship between mould release concentration, the time to mould stain and the release force.

As can be seen, increasing the release agent level results in reduced production time before mould cleaning is required. Simultaneously, however, the energy required for physical release from the mould is reduced. The physical release force shown here is that from an arbitrary laboratory test. By experience, values of 15 lb or less in this test can be expected to give satisfactory release in production moulding. A typical concentration of release agent would thus be in the region of 0·5%. Certain new synthetic release agents are now available

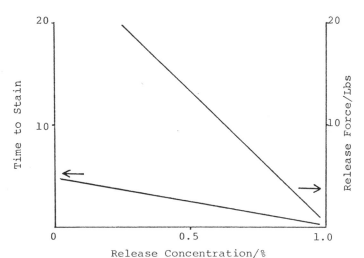

Fig. 5.8. The relationship between mould release concentration, time to stain and release force.

which can give improved staining performance and these are beginning to replace carnauba and candelilla wax in formulations offering improved stain-free performance.

5.3.5. Glass Transition Temperature (T_g) and Stress

With the mass production of 256 and 512K dynamic RAMS either occurring or imminent and the production of 1 megabit memory devices scheduled before the end of this decade, stress performance is becoming increasingly important. Even with considerable progress in the production of smaller and smaller device geometries individual die sizes are still increasing. Accompanying this is the requirement for more and more connections to be made to a chip, which means that 40 and 64 pin dual in-line plastic packages are now common. Stresses are caused by the thermal expansion coefficient mismatch between the plastic package and the silicon die and leadframe it encapsulates. They can affect the electrical properties of the device, causing cracking of passivation layers or even the chips themselves. In the most extreme cases stresses can even result in cracking of the encapsulant. The internal stress performance of an epoxide moulding compound is related to a number of factors, the important ones being the thermal expansion coefficient and the Young's modulus of the material. Interrelated with these properties is the glass transition temperature. Figure 5.9 shows the variation of internal stress with glass transition temperature for a typical epoxide moulding compound.

In order to obtain optimum stress performance in the past, some sacrifice in glass transition temperature has often been made. However, a high glass transition temperature is usually required to ensure good device reliability, particularly where high temperature operation or thermal cycling of devices is likely to occur. There has thus been a trade-off between stress and glass transition temperature. Recent developments have shown that for each compound there is a filler concentration where the thermal expansion coefficient and Young's modulus interrelationship can be optimised allowing low stress performance to be achieved whilst simultaneously giving higher glass transition temperatures. The concept of low stress moulding compounds is discussed further in Chapter 10.

5.3.6. Glass Transition Temperature (T_g) and Curing Conditions

For a given compound, the final glass transition temperature achieved is determined by the time and temperature of the cure and the level of

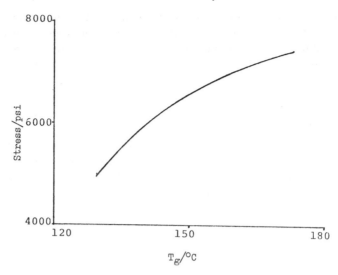

Fig. 5.9. Variation of internal stress with T_g.

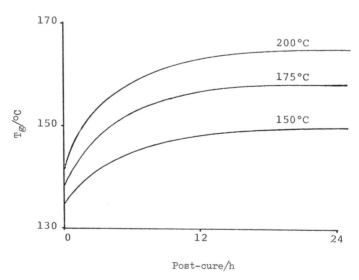

Fig. 5.10. Variation of T_g with cure time and temperature.

catalyst used in the formulation. When formulating a moulding compound, enough catalyst must be used to provide a good cure within the specified moulding cycle parameters. As can be seen in Fig. 5.10, T_g increases curvilinearly with time and temperature to different maximum levels depending upon the post-cure temperature.

The choice of post-cure conditions has a direct bearing upon the degree of crosslink density achieved and, in effect, on the overall properties of the cured compound. The zero point on Fig. 5.10 represents the as-moulded T_g before post-curing and thus variations in the permissible moulding conditions play a large part in determining the catalyst type and level to be selected for a given formulation. Whilst it would seem obvious to use additional catalyst where a curing schedule is limited, it should be borne in mind that device reliability usually decreases somewhat with increasing catalyst concentration. The use of a more reactive catalyst may also lead to problems with reduced shelf stability.

5.3.7. Shelf Stability

An ideal epoxide moulding compound would have indefinite shelf stability at room temperature. Unfortunately it is not possible to have a compound that will cure in 90 s at 180°C and yet will not react at room temperature. Shelf stability of epoxide moulding compounds is

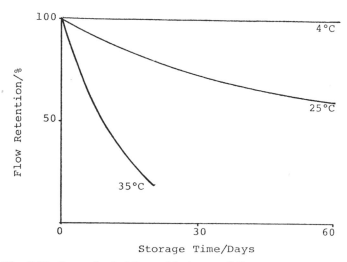

Fig. 5.11. Loss of spiral flow with time at different temperatures.

usually represented in terms of spiral flow retention, since as the compound reacts its spiral flow diminishes until it will be too low to fill a mould. It is normal to store epoxide moulding compounds at 4°C to increase their shelf-life and Fig. 5.11 shows the loss of spiral flow with time for a typical epoxide stored at several temperatures. At 4°C the flow retention is almost 100% even after 60 days.

At 25 and 35°C flow retention reduces significantly with time. There is obviously a trade-off between reactivity and shelf-life and it is usually true that the greater the reactivity of a compound the shorter its shelf-life will be. This should be borne in mind when making catalyst selections for a formulation. Ideally, a catalyst would be totally latent at room temperature, only becoming reactive at the moulding temperature.

5.3.8. Raw Material Purity and Cost

The purity of the materials used in the formulation of epoxide mould-ing compounds is of the utmost importance in determining device reliability since ionic impurities in a compound are known to cause corrosion failures. The purity of the constituents used in moulding compound formulations has been dramatically improved over the last few years, but materials with increased purity usually cost more than their less pure analogues. When formulating an epoxide moulding compound, the purity requirements need to be taken into account and considered against the likely cost penalties for improved purity. There is no point in using the highest purity raw materials for a compound designed to encapsulate reed relays or resistors, yet the utmost purity will be required for materials to encapsulate certain CMOS devices, where sensitivity to impurities may cause corrosion or parametric failures.

5.4. COMMON MOULDING PROBLEMS AND HOW TO AVOID THEM

Whilst the moulding of semiconductor devices in plastics using a trans-fer moulding technique appears to be a relatively straightforward operation, careful attention has to be given to a large number of parameters in order that properly moulded parts are obtained in acceptable yields. It is essential that the correct moulding compound is selected for a particular application, since a compound that is suitable for moulding diodes or transistors may be totally unacceptable for moulding VLSI circuits in high pin count packages. The moulding compound being used should be stored, prior to use, in the correct manner as recommended by the supplier. Normal storage procedures involve keeping the unopened drums of material at around 4°C and then allowing them to equilibrate with the ambient temperature prior to opening and use. Incorrect storage of material may lead to partial curing of the compound, with an accompanying loss of spiral flow and the possible inclusion of moisture. The use of a poorly designed mould or one that is badly worn may also lead to poor quality moulded parts with excessive flashing and voiding being common occurrences. Another important factor which directly influences the quality and yields of moulded parts is the use of moulding operatives with little or no training in the handling and moulding of the compounds.

Assuming that all the above factors have been taken into account, it is still possible that moulding problems may be encountered from time to time during operation. Some of the more common moulding problems and, fortunately, some of the most easily eradicated problems are now reviewed.

5.4.1. Flashing and Resin Bleed

The term 'flashing' is applied to moulding compound that has flowed into unwanted areas, usually in the parting line region, although with worn knockout pins flashing may also occur forming so-called 'chimneys' on a moulded part. Excessive flashing leads to parts which, at worst, are totally unacceptable or in less severe cases require additional cleaning time. Cleaning will also be required to remove any flash remaining in the mould, since any residual flash may prevent the mould closing properly during the next cycle leading to a repetition of the problem.

A typical example of very bad flashing is shown in Fig. 5.12 on a moulded 14 pin leadframe assembly.

Resin bleed refers to the separation of the organic components (resins) of the moulding compound from the inorganic components (fillers). Resin bleed is usually a precursor to flashing since material is flowing into areas where it is not supposed to, the size of the gap along which flow occurs being too small to allow the filler particles to migrate. Resin bleed adheres tenaciously to the surfaces it comes into contact with and this leads to problems with leadframed devices in which the leadframe has to be plated or soldered in a later operation. There are a

Fig. 5.12. Flash on moulded devices.

number of causes of flash and bleed, the principal ones being:

(1) *High transfer pressures.* The use of an excessively high transfer pressure may result in flash and bleed along the parting line area. The transfer pressures recommended by the moulding compound manufacturer should be used and it is essential that the clamping pressure of the mould is sufficient. As a guideline, the clamping pressure should be approximately five times greater than the transfer pressure.

(2) *Worn moulds.* As with most mechanical equipment, continued use leads to wear and, in the case of a mould, the parting line area becomes worn allowing flashing and bleeding to occur. As the wear increases it will eventually be necessary to replace or at least retool the mould. It may be possible, as a palliative measure, to prolong the use of worn moulds by using moulding compounds with shorter spiral flows and/or increased melt viscosity. This can be accepted for use with TAB and bump chip type assemblies but with standard wire-bonded leadframes, wire bond sweep will eventually become a problem giving reduced yields.

(3) *External release agents.* Epoxide moulding compounds are generally formulated to incorporate release agents and additional external release agents are largely unnecessary. The regular use of an external release agent on the mould surface may act as a

plasticising agent for the moulding compound. This results in additional flashing and, more particularly, bleeding of the resin. The use of external release agents should therefore be avoided, or at least kept to a minimum, whenever possible. It is also possible that the internal and external release agents may be incompatible.

(4) *Out of specification inserts.* If lead-frames or device leads are not produced within a rigidly specified thickness range, flashing and bleeding can be exacerbated. Parts thicker than an acceptable tolerance will prevent the mould closing properly, with flash and bleed occurring around the standard thickness parts. Conversely, thin parts will allow flash and bleed around themselves, because, even though the mould is closed properly, there may be a gap between the mould and the thinner leads or leadframes.

(5) *Unsuitable moulding compound.* The use of a compound with an abnormally low melt viscosity may well lead to bleed problems and possibly flashing. Filler particle size is also important because flash and bleed increase with decreasing filler size. These two parameters are controlled by the moulding supplier and with persistent problems the manufacturer should be consulted for help with flash and bleed control.

5.4.2. Knit Lines and Incomplete Fill

Knit lines and incomplete fill are said to occur when the mould cavity remains only partially filled and/or the moulding is not homogeneous, with grooves appearing in the pieceparts. Knit lines are at best cosmetically objectionable and at worst sites for the entrapment of moisture and various ionic contaminants that can lead to early failure of a device due to corrosion. A typical example of non-filling is shown in Fig. 5.13. The more common causes of incomplete fill and knit lines are:

(1) *Insufficient or excessive amounts of moulding compound.* In a normal moulding operation preforms of the moulding compound are used having a specific weight in order to completely fill all the cavities of the mould and to give a cull of suitable thickness to allow the pressure of the ram to be transferred to the material at all times. Using less than the required amount to fill the mould will obviously result in incomplete fill. The usual reason for the inadvertent use of less than the ideal amount of material is variation in preform density, where the use of several preforms

Martin T. Goosey

Fig. 5.13. Incomplete fill.

of low density will result in a significant loss of material. Less obviously, the use of too much material may also lead to incomplete fill. This occurs because the larger than normal mass of material does not liquefy sufficiently, the melt viscosity not reaching the required minimum. By paying attention to the cull thickness, it is relatively straightforward to determine if too much material is being used. The use of too much material will also lead to excessive waste and increased production costs.

(2) *Incorrect moulding pressure and temperature.* Moulding temperatures and transfer pressures should be maintained well within the range set by the moulding compound manufacturer for the particular material in use. Whilst the maintenance of these correct temperatures and pressures would appear to be fairly straightforward, low transfer pressure or low temperature is all too often the cause of incomplete fill. On many machines, the only accurate way of setting the transfer pressure is to use a force gauge which the ram can act upon directly to allow an exact value of the transfer pressure to be determined. A low moulding temperature may be used inadvertently because of reliance upon inaccurate temperature gauges on a press or because of inoperative heating elements within the mould that give cool areas. A too low moulding temperature may prevent the moulding compound from achieving the minimum melt

viscosity required for complete mould filling. Conversely a too high moulding temperature may accelerate the reaction rate of the epoxide leading to premature gelation of the compound before all the cavities have been filled.

(3) *Transfer speed.* The transfer speed or fill time is also very important in determining whether complete fill of all the cavities is achieved. A too rapid transfer can cause heating of the compound in the restricted area of the gates, which may lead to premature gelation or blockage of the gate. A too slow transfer will mean that the compound gels before the transfer process is complete. Both long and short transfer speeds can thus lead to incomplete fill. As a general rule for average size moulds, a transfer time of between 15 and 20 s would normally be acceptable.

(4) *Preheating.* Dielectric preheating of parts is usually essential to achieve good quality moulded parts. For any given application and set of moulding parameters there will be a narrow temperature range from which the preheated preform temperature should not vary. Too low a temperature may lead to incomplete fill because the melt viscosity will not achieve the required minimum necessary to completely fill the cavities. Too high a preheat temperature may lead to premature curing and gelation of the compound before all the cavities are filled.

(5) *Incorrect or short flow.* On any new moulding application the flow requirements of the required compound must be established within the limitations imposed by the moulding operation and the suppliers specification. With all the other important parameters fixed, an application requiring a compound with a spiral flow of 35 in will have incomplete fill problems if a compound with a spiral flow of 25 in is used. Where a material with the correct spiral flow has been selected, there is still the possibility of incomplete fill if the compound has been exposed to high temperatures during shipping or storage that have allowed the compound to partially cure thereby reducing the spiral flow. It is obviously important that a good correlation has been achieved between the spiral flows as determined by the customer and the supplier.

(6) *Inadequate venting.* Although epoxide moulding compounds cure by an addition reaction, without the evolution of volatile reaction products, vents are still required in a mould to prevent

the entrapment of air, absorbed moisture or any components evolving from the component to be encapsulated. The vents are usually situated opposite the gates and should be deep enough to allow the evolution of volatiles without the build up of a back pressure. Inadequate venting can lead to air entrapment, incomplete fills and knit lines. Particular attention has to be given to cleanliness in the vent areas since the build up of wax or resin bleed may lead to their partial or complete blockage.

(7) *Gate blockage.* Incomplete fill is also sometimes caused when gates become partially or completely blocked, preventing or impeding the flow of the moulding compound into the cavity. Usual causes of gate blockage are oversized filler particles and partially cured moulding compound. The problem is particularly likely to occur in semiconductor applications with small-outline (SO) type packages where gates less than 0·010 in deep are becoming common. The presence of partially cured particles of moulding compound usually arises because of cleanliness problems in the mould or in the preforming operation.

(8) *Non-heated inserts.* It is normal practice to heat the parts to be moulded before they are inserted into the mould, in order that they are at the same temperature as the mould and moulding compound. This is particularly important when large pieceparts are being moulded, because compound entering a cavity and impinging upon a cool insert may experience an increase in viscosity that can inhibit the flow, again resulting in incomplete fill.

(9) *Worn equipment/insufficient clamping pressures.* The use of worn moulds, transfer pots and rams, or the use of insufficient clamping pressure can lead to the excessive flashing of a compound into areas where material is not required. Moulding compound consumed in flash is not available for filling mould cavities and thus incomplete fill may again be the result.

5.4.3. Blisters and Voids

Blisters, as their name implies, appear as eruptions in the moulding compound surface, and are due to the expansion of underlying trapped gases or volatiles when the moulding pressure is relieved. Voids occur as small irregular imperfections in the surface of the moulding compound, although they also sometimes appear within the bulk of the compound. The usual causes of blisters and voids are:

(1) *Excessive volatiles.* The usual problem of volatiles is caused by an excessive free moisture content within the moulding compound which is unable to be vented during the moulding operation. Excess moisture usually finds its way into a moulding compound through poor storage procedures. Moulding compounds are typically stored at temperatures around 4°C and if a drum is opened before its temperature has equilibrated with room temperature there is every likelihood that moisture will condense out of the atmosphere onto the cooler moulding compound. This is particularly likely if the operating environment is one of relatively high temperature and humidity. Correct storage procedures would include allowing the material to warm up to room temperature before the drum was opened and avoiding excessive atmospheric exposure to any open drums.

(2) *Rapid transfer.* If the transfer speed is not adequately controlled a too rapid transfer may occur, leading to turbulent or 'jet-like' flow of the material into the mould cavities. This can cause vents to be blocked before all the unwanted volatiles have escaped from the cavity. Too fast a transfer may also lead to excessive heating of the compound and premature gelation which can cause void formation.

(3) *Inadequate venting.* Just as inadequate venting can lead to non-filling of cavities, a precursor to this situation may be the formation of voids and blisters.

(4) *Excessive moulding temperature (blisters).* Epoxide moulding compounds are usually rigid enough at the end of a moulding cycle to withstand any internal pressures caused by small amounts of trapped air or volatiles. However, if an excessive moulding temperature is used, the epoxide may become sufficiently elastic for more pressure relief to occur, often resulting in blister formation when the moulding pressure is removed. At even higher temperatures thermolysis of the moulding compound components can lead to further volatile evolution with subsequent blister formation.

(5) *Melt viscosity (voids).* The melt viscosity during a moulding operation is one of the most important parameters that influence a moulding compound's mouldability. An abnormally low melt viscosity can give uncontrolled cavity filling leading to the en-

trapment of volatiles and gases, with the subsequent formation of internal or external voids. As with the non-fill situation, the use of non-heated inserts may lead to an increase in melt viscosity which may, in less serious cases, give voids in the moulding.

5.4.4. Staining of Moulds and Poor Cosmetic Appearance

The staining of a mould usually refers to a discoloration of the mould surfaces by the deposition of unwanted residues from the moulding compound. The staining is typically brown in colour and represents either the colour of the residue or the colour of its oxidation product. Typical causes of these residues are minute amounts of natural impurities found within some of the natural waxes used in the moulding compound formulation as release agents. The effect is typically seen with epoxide novolac based moulding compounds and continued staining or build up of residues eventually leads to cosmetically objectionable parts which may exhibit markability problems. A typical cause of staining and poor cosmetic appearance is excessive temperatures. The use of excessive moulding temperatures leads to a decrease in the operating time before the degree of discoloration becomes objectionable, because of accelerated oxidation of the waxes. By selecting a lower moulding temperature the staining can be reduced, but in doing so the cure time may need to be extended to avoid undercure of the devices and possible problems with sticking.

5.4.5. Sticking in the Mould

Mould sticking refers to the incomplete removal of parts from a mould due to adhesion between the moulding compound and the metal surfaces of the cavity. In less severe cases, a part may be released largely intact, but its surface will have imperfections in it and there will be residual moulding compound adhering to the mould surfaces. More serious sticking could actually result in the parts being physically cracked or torn apart, as can be seen in Fig. 5.14. In both cases, extensive cleaning of the mould will be required between consecutive moulding cycles, reducing productivity and yields unacceptably. Typical causes of sticking are:

(1) *Undercure of the moulding compound.* Undercured moulding compounds tend to adhere well to mould surfaces because they have not achieved the crosslinking density, cure shrinkage and hot

Fig. 5.14. Parts cracked because of a sticking problem.

hardness of the fully cured compound. The two main reasons for this are the use of too low a moulding temperature and too short a cure time. Each individual compound will have a specific minimum mould cycle time/temperature relationship below which problems can occur because of undercure of the material. The moulding conditions specified by the compound manufacturer will usually allow a safety margin but, as a general rule, if it is necessary to use a lower than recommended moulding temperature the mould cycle time should be increased. Should mould cycle time be important, a long cycle time can be reduced in stages until the point at which sticking occurs is reached, and then increased slightly again to avoid the problem.

(2) *Incompatible release agent.* Even though most moulding compounds are formulated with an internal mould release, additional external mould release is often used, particularly when breaking in a new mould or when beginning work on a little-used press. When it is necessary to use an external release agent, every effort should be made to ensure that it is compatible with the moulding compound in use since certain external releases may actually interfere with the material's curing reaction leading to sticking as in (1) above. The use of external release agents should be kept to an absolute minimum and, where there is any

doubt regarding compatibility, the compound supplier should be contacted.

(3) *Dry moulds.* One particular case where the application of an external mould release agent may indeed be necessary to prevent sticking is where a mould has dried out. With continued use, the mould release agent permeates into the micropores of the mould, promoting a weak boundary layer between the metal and the moulding compound. However, if a mould is left at temperature for excessive periods without being used this film of release agent will lose its effectiveness because of thermal degradation. Such a dry mould will lead to very bad sticking until the release film has been re-deposited. To assist this process small amounts of external release agent can be sprayed into the cavities and runners, any excess being wiped off with a lint-free rag. Usually the external release used should be the same material as the internal release and a discussion with the material supplier will often help. In addition to this external lubrication, the first shot moulded should be given a considerably longer cycle time in order for the release from the compound to be allowed to 'soak in'.

(4) *Poor mould design.* Draft angles are usually machined into mould cavities and, combined with the cure shrinkage of the compound, facilitate easy release from the cavity. Sticking may occur in moulds with shallow or no draft angles when compounds with low shrinkage values are moulded and this is particularly important with the latest generation of low stress compounds, where draft angles of 8–12° are usually recommended. Good mould design also dictates the use of knock-out pins in the runners and cavities and the presence of these can often make the difference between sticking and good release.

5.5. EVALUATION OF EPOXIDE-BASED TRANSFER MOULDING COMPOUNDS

In order for a moulding compound to be suitable for semiconductor encapsulation, it must have a number of special characteristics which need to be well defined and controlled within certain critical limits. This section discusses some of the most important properties that affect mouldability of epoxide moulding compounds and details methods by which an end user can evaluate them.

5.5.1. Hot Plate Cure Time

The hot plate cure time of a moulding compound is a very important parameter because it gives a direct indication of how long the compound will take to gel at a given temperature. This value gives a good guide to establishing the mould cycle time of a moulding operation and also the time available to transfer the liquefied epoxide from the pot into the individual cavities. As its name suggests, the hot plate cure time is determined simply by placing a small amount (2–3 g) of the moulding compound under investigation onto the surface of a hot plate at the required moulding temperature and kneading it backwards and forwards with a wooden spatula until it gels. With practice, the gel time can be determined for a moulding compound with an accuracy of better than ±1 s.

5.5.2. Spiral Flow

The spiral flow is the distance a material will flow in a specific type of mould under a given set of conditions. The spiral flow of a compound is obviously very important since it should always be longer than the maximum distance the compound has to travel in the press. If a compound is exposed to any temperature excursions or allowed to sit at room temperature for long periods its spiral flow will decrease and so this method is a good way of assuring the material is still within specified limits.

The test is performed with a transfer pressure of 1000 psi, which is measured directly at the ram face with a Dillon type force gauge. Platens should be set to the required temperature and controlled to within ±5°F. A commonly used temperature is 350°F (177°C). The material is moulded as a powder and a charge is used sufficient to give a cull thickness of between 0·12 and 0·14 in. Some trial shots will usually be required to establish the required charge size for a given compound to give a cull thickness within the specified limits. The spiral flow is recorded to the nearest $\frac{1}{4}$ in at the point of farthest flow where the spiral is continuous, and a mean value of at least two shots, which must agree to within ±3% is reported. A typical cull and spiral is shown in Fig. 5.15.

5.5.3. Viscosity and Mouldability

The viscosity versus time characteristics of a compound during the transfer moulding operation are critically important factors in the achievement of successfully moulded parts. The fluidity of the material

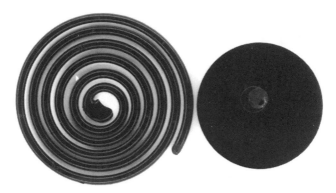

Fig. 5.15. Typical cull and spiral from a spiral flow test.

under the effects of both temperature and pressure must be controlled within well-defined limits, so that stress-free parts are moulded, mould cavities fill properly and problems such as wire bond sweep do not occur. There are various methods for the quantitative determination of flow and viscosity of moulding compounds but the principal three are: the ram follower/spiral flow test, the Koka flow test, and the torque rheometer (Brabender plastograph) test.

5.5.3.1. Ram Follower/Spiral Flow Test

This is one of the most common methods for evaluating flow and viscosity, and utilises the spiral flow mould mentioned above. The object of this test is to measure the movement of the transfer ram with time during a moulding cycle using the spiral flow mould at a specified temperature and pressure (usually those required for the spiral flow test). By recording the movement of the ram, the displacement of the moulding compound is effectively monitored. It is possible to obtain a direct read-out of ram travel against time if the ram is connected to a chart recorder via the appropriate interfacing circuitry and the resulting plots can be directly compared from compound to compound or batch to batch.

5.5.3.2. Koka Flow Test

This method utilises an extrusion process, whereby the material under investigation is forced through a nozzle to impinge upon a plunger. The movement of the plunger is measured against time in a similar way to the ram follower method and again the resultant curve yields information on the gel time and viscosity of the material.

5.5.3.3. *Torque Rheometer (Brabender Plastograph) Test*

This method utilises a torque rheometer, of which the Brabender plastograph is one of the most commonly used, to monitor viscosity of the material under examination against time. This is achieved by introducing the compound into a heated mixing chamber where it interacts with two activated rotors. These rotors are connected to a torque meter and the changes in rotation torque are measured against time. Upon introduction, the torque reading will be high, but as the compound heats up and melts its viscosity drops and the torque reading falls. However, at the elevated temperature of the chamber the curing reaction occurs and so the viscosity, and thus the torque reading, begins to increase until the compound is gelled. From this type of experiment the minimum melt viscosity, the reactivity and the gel time of a compound can all be evaluated. Typical results for this type of test performed at various temperatures are shown in Fig. 5.16.

With all of these three tests it should be remembered that the behaviour of the compound will be different to that occurring in a normal device moulding operation and some allowances will need to

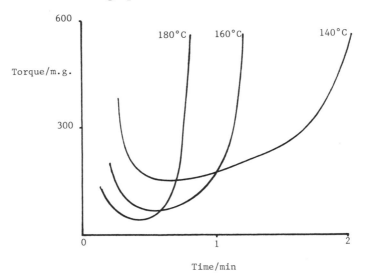

Fig. 5.16. Typical torque versus time curves for a moulding compound at different temperatures.

be made to take this into account, particularly since the flow of a compound into mould cavities will be highly mould design dependent.

5.5.4. Hot Hardness

The hot hardness of a moulding compound at the end of a moulding cycle is important because, if the moulded surface hardness is insufficient, cracking of parts can occur during ejection. The hardness of a moulding compound represents a collection of various interrelated properties rather than a single physical property of a material. The hot hardness test is usually performed on the cull from a spiral flow test. A reading is taken on the cull as soon as the press opens using a precalibrated Shore D durometer which is pressed onto the cull until it is seated flat. The Shore D hardness number can then be read directly from the durometer. By making measurements of hot hardness for various mould cycle times it is possible to record the increasing hardness of a compound as it becomes fully cured. A typical example is given below:

	Mould cycle time (s)			
	60	75	90	105
Hot hardness at 180°C	74	77	80	85
Hot hardness at 160°C	48	50	69	76

5.5.5. Flash and Bleed Characteristics

The flash and bleed characteristics of a moulding compound can be evaluated relatively easily using a specially prepared flash mould. This mould essentially consists of a series of channels of varying depths, usually emanating from a central cull area. The shallowest channel would have dimensions as little as 5 μm and the other channels would have increasing depths such as 10, 25, 50 and 100 μm. Depths of 25 μm or less would be expected to allow only resin bleed, whilst at depths greater than 25 μm flashing would also occur. Materials are evaluated by measuring the lengths that the flash or bleed has travelled along a particular channel. For a given compound, the flash and bleed characteristics will obviously vary from moulding operation to operation and will be dependent on a large number of factors. It is therefore important to ensure that any material evaluations are performed on the same mould under identical operating conditions.

5.5.6. Mould Release

There is no absolute method for quantitatively determining the mould release of a compound and, in practice, measurements are often performed using an arbitrary mould. A typical example would be to mould a tapered slug of material and then measure the force required to free the taper from the mould. The differences in the values obtained for various materials can then be related to their release performance in an actual moulding application. Experience with a mould under carefully controlled conditions will often allow a release force value to be determined below which release performance on the application under investigation can be considered acceptable.

ACKNOWLEDGEMENTS

I would like to acknowledge the contribution made to parts of this chapter by the technical staff of the Dynachem Corporation's Polyset Division. Thanks are also due to Paul Litke of the Morton Chemical Company, Woodstock, Illinois, USA, and to Kenneth Cluckey and Andrew Vavra of MTI Corporation, Ivyland, Pennsylvania, USA, for supplying photographs of equipment and devices.

Chapter 6

PLASTIC ENCAPSULATED DEVICE RELIABILITY

R. P. MERRETT

British Telecom Research Laboratories, Ipswich, UK

6.1. INTRODUCTION

The introduction of planar technology in 1959 dramatically reduced the price of semiconductor components and, as a result, encapsulation costs became a significant proportion of the total. At that time most components were hermetically packaged, but by 1962 the pressures to make some high-volume production items less expensive led to the development of plastic encapsulation. Since then the use of plastic encapsulation has increased to the extent that over 90% of all semiconductor products are now supplied in this form.

Plastic components were accepted at the outset by the consumer industry virtually without reservation because of their principal attraction of low cost. However, the computer, telecommunications and military sectors have long had strong reliability reservations and it is only in very recent years that economic pressures, which fortuitously coincided with significant improvements in quality and reliability, have led to the acceptance of plastic components by these industries. The only areas where plastic encapsulation is unlikely to be used are for the high reliability applications, such as military and aerospace and for undersea repeaters, where the need for guaranteed reliability takes precedence over cost.

It was in 1966 that the first paper raising doubts on the quality and reliability of plastic encapsulated components was published and this was the beginning of the many hundreds of papers which continue unabated to this day. By 1968 the key factors which affect reliability

173

had been correctly identified (i.e., the quality of the plastic and the passivation layer), but adequately reliable components were not generally available until the early 1980s. The prime reason for this sluggishness was that changes in encapsulation technology were only made to contain or reduce production costs, and these changes did not necessarily improve quality and reliability. Specific attempts to improve reliability were generally only introduced when there was sufficient pressure from users, and then only when the production costs were not unduly compromised.

This chapter is written from the standpoint of a discerning user who is concerned with procuring, at minimum cost, plastic components of adequate reliability. After briefly discussing the relevant aspects of plastic encapsulation, and the techniques used to quantify reliability performance, the physical principles of the four main package-related failure mechanisms will be discussed and particular emphasis will be placed on the methods of reliability assessment.

References will usually only be given when the results of recent investigations are introduced. Comprehensive bibliographies on the reliability of plastic encapsulated components and on the major failure mechanisms are given elsewhere.[1-3]

6.2. THE TECHNOLOGY AND FAILURE MECHANISMS

6.2.1. Package Types

A cross-section through an example of one of the commonest forms of plastic package is illustrated in Fig. 6.1(a). The first stage of construction is to attach the semiconductor die to a metal leadframe. After die attach, the electrical connections between the bond pads on the IC and the leadframe are made using fine wire, and the assembly is then encapsulated in a thermosetting resin. The dual-in-line form of this package is the most popular, but increasing use is now being made of the surface mounting formats, namely the chip carrier and the small-outline (SO) packages. Although these various types of package differ in shape, and in the configuration of the external leads, they are all made in the same way. Accordingly there is no need to make any distinction between them when discussing failure mechanisms.

The only other form of plastic encapsulation is illustrated in Fig. 6.1(b) which shows a layer of plastic covering a die that is either

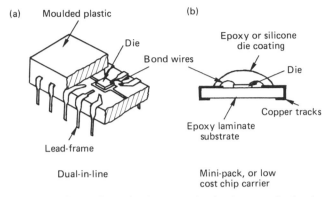

Fig. 6.1. Cross-sections through the two basic forms of plastic package: (a) dual-in-line; (b) mini-pack or low cost chip carrier.

mounted onto an epoxy laminate chip carrier or is directly mounted onto a printed circuit board (as in some cameras and watches). Various types of plastic, which are often referred to as junction coatings, have been used for this type of encapsulation but room-temperature vulcanising (RTV) silicone now appears to be the most popular. Potentially, this type of encapsulation could have a very low cost in volume production, but it has so far failed to attract widespread interest.

6.2.2. Wire Bonding

During transfer moulding, the bond wires connecting the semiconductor die and the leadframe are subjected to stresses caused by the inrush of plastic into the mould. To ensure that wire-bond integrity is maintained during the moulding process, the bonding operation must be optimised to a higher degree than for hermetic encapsulation. Typically, 1·25 mil diameter gold wire and thermocompression ball-bonding are used for plastic encapsulation, whereas 1·0 mil diameter aluminium wire and ultrasonic wedge-bonding are used for hermetic encapsulation.

Bond wire failures which occur during service life, as opposed to the moulding operation, are the result of: (i) weakening of the bond due to the growth of intermetallic compounds at the interface between the gold wire and aluminium die metallisation; and/or (ii) thermally induced stresses caused by the difference between the coefficient of expansion of the plastic and the leadframe.

6.2.3. Leadframe and Die Attach

For many plastic encapsulated ICs, there is sufficient power dissipation to raise the temperature of the die significantly above that of the ambient. This increase in temperature enhances the rate of failure due to: (i) parasitic transistors and conducting channels caused by the accumulation of ions on the surface of the die; (ii) electromigration of the metallisation tracks; and (iii) intermetallic growth at the bond pads. In addition, the thermal stresses introduced by switching an IC on and off can cause the bond wire failures mentioned in the previous section.

The rise in temperature of an IC encapsulated in plastic is primarily determined by the thermal conductivity of the leadframe material, and is generally about twice that of its hermetically packaged equivalent. Until recently, the leadframe material has been exclusively a nickel–iron alloy (Alloy 42), and for a 14 lead dual-in-line package the thermal resistance (i.e., rise in die temperature per unit power dissipated) is about 120°C/W. In order to cope with the increasing power dissipation of ICs, high copper content alloys (OLIN) have been introduced to reduce the thermal resistance by about 50%.

Changes have also occurred in the type of die attach materials employed. At one stage gold–silicon eutectic was used exclusively, but it has the disadvantages of being difficult to use for large dies, it is expensive, and the high temperatures required (415°C) can cause oxidation of copper alloy leadframes. Accordingly it is being replaced in many cases by conductive (silver-loaded) epoxies or polyimide.

6.2.4. Thermosetting Plastics

The moulding compounds used for encapsulation are a mix of a resin (either epoxy or silicone), a hardener (for epoxies either a novolac or an anhydride), an inorganic filler (such as silica or alumina) together with a mould-release agent, a flame retardant and proprietary ingredients used to optimise some characteristics of the moulding compound.

Epoxy novolac formulations dominate the market and are used for over 90% of ICs. Silicones are used almost exclusively for discrete power devices because of their better thermal stability, and epoxy anhydrides are sometimes used as a cheap replacement for silicones. The reasons for the preference of epoxy novolacs over epoxy anhydrides for all but higher temperature applications are very practical manufacturing ones, namely a better shelf-life and ease of moulding. This latter requirement is absolutely crucial to the semiconductor component manufacturer who is concerned with achieving high yields.

All plastics are permeable to moisture, and this can lead to failures caused by: (i) electrochemical corrosion of aluminium metallisation; (ii) dendritic growths between gold metallisation tracks; and (iii) the corrosion of gold if suitable complexing agents are present. The rate of degradation is affected by many factors but two of the most important are the amount of ionic contamination which can be leached from the plastic and the quality of the passivation layer which is now present on the surface of all ICs.

6.2.5. Passivation

Die passivation layers were first introduced to protect the metallisation of integrated circuits from mechanical damage during assembly. However, it soon became apparent that the passivation layer was a crucial factor in determining the failure rate in moist ambients. Originally, pure chemical vapour deposition (CVD) silicon dioxide was used, but later phosphorus was added to relieve strain in the layer and thus prevent cracking and loss of adhesion. There is also interest in, and limited use of, other forms of passivation such as silicon nitride, oxynitride and polyimide.

6.3. RELIABILITY ASSESSMENT TECHNIQUES

6.3.1. Accelerated Lifetests

Most failure mechanisms involve the diffusion of atoms and therefore have an Arrhenius-type time dependence

$$\text{Reaction rate} \propto \exp\left(\frac{-\Phi}{kT}\right) \tag{6.1}$$

where T is the temperature (K), k is Boltzmann's constant and Φ is the activation energy. The rate of degradation can thus be accelerated by increasing the temperature, and it is thereby possible to compress the service life of components into the scale of a laboratory experiment. These tests, which are often referred to as 'stress tests', are the basis of most reliability studies.

In order to determine the value of Φ it is necessary to subject at least two groups of specimens to lifetests at different temperatures and to measure a parameter, such as the mean time before failure (MTBF), related to the rate of degradation. For most thermally induced failure mechanisms, the value of Φ will lie in the range $0\cdot45$–$1\cdot2$ eV.

Having obtained the activation energy it is then possible to use the

MTBF at one of the test temperatures to determine the MTBF corresponding to the normal operating temperatures. The ratio of the two times is known as the acceleration factor, A, and is given by

$$A = \exp\left[1 \cdot 163 \times 10^{-4} \Phi \left(\frac{1}{T_0} - \frac{1}{T_s}\right)\right] \tag{6.2}$$

where T_s and T_0 are the stress and the normal operating temperatures (K), respectively.

For some failure mechanisms, factors such as track current supply voltage, and relative humidity can also influence the rate of degradation, although the functional dependence is rarely as simple as that for temperature. Lifetests can thus be based on various combinations of stress conditions, but it must always be recognised that the conditions chosen may be severe enough to introduce extraneous failure mechanisms which would not normally occur under normal service conditions. It is therefore necessary to determine the cause of failure on lifetests and to examine critically the results obtained.

6.3.2. Failure Analysis
To determine the cause of failure, it is usually necessary to remove enough plastic to expose the die. A variety of decapsulation techniques has been advocated but the acid-jet technique is the most effective. The package is placed in a jig which exposes the plastic immediately above the die to a jet of fuming sulphuric acid (at 240–300°C). The plastic is rapidly etched by the jet and, provided care is taken to ensure that the acid does not come into contact with the die for more than a few seconds, the amount of damage to the latter is minimal.

6.3.3. Reliability Mathematics
The time to failure of specimens subjected to an accelerated lifetest is a random variable which is usually distributed according to a log–normal cumulative probability function (CPF). Once the shape and the values of the parameters characterising the CPF have been determined, it is possible to predict the failure rate of the sampled population as a function of the time in service.[4]

In order to determine which type of CPF is appropriate for a particular group of components, it is necessary to continue lifetests until at least about 50% of the specimens have failed. The statistical tests which can then be applied to determine the type and characteristic parameters of the CPF are described in standard texts. Fortunately, the results of lifetests on semiconductor components can usually be described by a log–normal CPF so, for present purposes, attention can be restricted to the simple graphical solution which is appropriate to this particular case.

For most lifetests it is impractical to monitor the performance of the specimens while the test is running, so it is not usually possible to detect failures at the instant they occur. Instead, the lifetests are interrupted at frequent intervals to permit assessment of the performance of the specimens, and the number of failures at stages during the test is thereby determined. There is then a choice of several equations which can be used to calculate a statistically unbiased estimate of the cumulative probability of failures. One of the most popular of these equations is

$$\text{Cumulative probability} = 100 \frac{(i - 0\cdot5)}{n} \% \qquad (6.3)$$

where i is the cumulative number of failures and n is the number of specimens on test, i.e. the sample size. If the results thus obtained are a reasonable fit to a straight line when they are plotted on log–normal probability paper, it can be assumed that the data come from a population having a log–normal CPF. The line can then be used to obtain a graphical estimate of the median lifetime (t_M) and the standard deviation (σ). For the log–normal distribution, the latter is given by the expression

$$\sigma = \ln(t_M/t_{16}) \qquad (6.4)$$

where t_{16} is the time at which 16% of the population has failed. Having obtained these two parameters, and if the acceleration factor A is known, it is then possible to predict what the failure rate $\lambda(t)$ of the components would be after a time, t, in service. This failure rate is defined such that $\lambda(t)\,dt$ is the probability of a specified component failing in the interval between t and $t+dt$. If both t_M and σ are expressed in hours, the failure rate can be calculated by either referring to published tables or using the expression

$$\lambda(t) = \frac{\exp(-z^2)}{\text{erfc}(z)} \sqrt{\left(\frac{2}{\pi t^2 \sigma^2}\right)} \text{ FIT} \qquad (6.5)$$

with

$$z = \frac{1}{\sigma\sqrt{2}} \ln(t/t_M)$$

R. P. Merrett

where one FIT (from 'failure unit') corresponds to a 10^{-9} probability of a specific device failing in a 1 h period. For medium-scale ICs, a failure rate of 30 FITs is acceptable and, for example, in a small telephone exchange containing about 10 000 such components this would correspond to three failures every year.

Two examples of the log–normal probability plots obtained from high temperature/high humidity tests on some plastic encapsulated components are given in Fig. 6.2, and the predictions for service failure rate are shown in Fig. 6.3. It should be noted that for the log–normal distribution the failure rate gradually rises during the service life, although it will eventuallly go through a maximum when about half the components have failed.

With most lifetests, practical difficulties and cost limit the size of the sample to less than 100, and sometimes to as low as 20. Thus it is important to consider the extent to which small sample sizes can affect the accuracy of the predictions of service performance. The derivation of confidence limits is outside the scope of this discussion but is covered in most statistical texts.

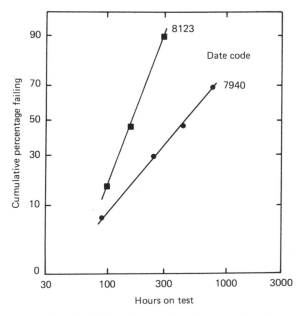

Fig. 6.2. Log–normal probability plots for the time to failure for two groups of CMOS dual-in-line ICs subjected to a lifetest in a 110°C/90% RH environment.

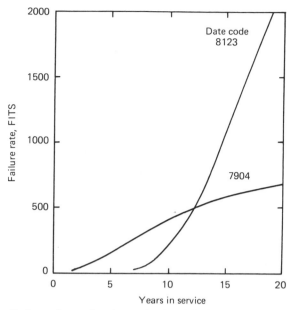

Fig. 6.3. Predictions, from the data of Fig. 6.2, for the failure rates of two groups of CMOS dual-in-line ICs in an uncontrolled UK environment (i.e., an average of 10°C/80% RH).

6.3.4. Test Philosophy

The reliability of a particular type of component will vary from production batch to batch, but it would be uneconomic to perform the variety of lifetests needed to predict service performance on every batch. However, by performing such a series of tests before a particular type of component is used in service, it is possible to identify and characterise the failure mechanisms, and to design simpler tests which can be regularly used to assess whether future production batches are likely to meet the required reliability standards. These simpler tests are mostly based on a specification of the maximum permitted failure fraction after a particular stress test.

6.4. THERMALLY INDUCED FAILURES

The three main thermally induced failure mechanisms of plastic encapsulated ICs are:

(1) The formation of brittle intermetallic compounds at the interface

between aluminium metallisation and gold bond wires; this failure mode is often referred to as 'purple plague' in recognition of the colour of one of the intermetallic phases. Gold bond wire is always used with plastic encapsulation, because the ability of a gold ball bond to withstand the stresses experienced during moulding is considerably greater than that of the bonds formed with aluminium wire. (The latter is used for most hermetically encapsulated devices, and in this case intermetallic formation will only occur when these wires are attached to a gold plated lead-through; this reliability hazard can be minimised by limiting the thickness of the gold plate and thereby preventing the formation of the gold-rich, brittle phase.)

(2) Parametric changes and a reduction in breakdown voltage induced by the accumulation of charge near the silicon surface. The charges can be either hot electrons which are emitted from active devices and then trapped in dielectric layers, or ionic impurities which, under the influence of electric fields, migrate through dielectric layers and/or migrate along the interface between such layers.

(3) Electromigration of the metallisation tracks and of silicon at the contact windows. Both metal migration and silicon migration are primarily functions of current density but there is also a dependence upon temperature.

Although these failure mechanisms are common to all forms of packaging, there are two factors which can make the failure rate of a plastic encapsulated IC greater than that of its hermetically encapsulated equivalent. Firstly, the higher thermal resistance of a plastic encapsulated IC results in a higher die temperature for a given power dissipation. Secondly, there is a risk that ionic species in the moulding material will migrate to the die and cause degradation. The influence of the ionic content of the plastic on thermally induced failure mechanisms has not been comprehensively studied but it is reasonable to postulate that such ions may be involved in some of the failures caused by charge accumulation, and there have been reports that 'purple plague' is aggravated by the presence of bromide flame retardants and chlorine impurities in moulding compounds.

In order to assess the susceptibility of a plastic encapsulated component to thermally induced stress, there is no alternative but to perform lifetests at temperatures which will give a substantial acceleration of the failure rate relative to service conditions. At one stage, reverse bias was applied during lifetests in order to further enhance the acceleration. However, because the effect of reverse bias is hard to quantify, and because it is impractical to use with many ICs, it is now common practice to bias in a manner which is representative of the expected operating conditions. It is also worth noting that if the lifetests are being performed solely to assess the risk of failures due to intermetallic formation there is no need to apply bias.

The degree of acceleration achieved by using a particular stress temperature depends on the activation energy of the failure mechanism to which the components are most susceptible. Ideally, the value of this activation energy should be determined by performing lifetests at two temperatures, but such a procedure is usually too costly and time consuming for either production line monitoring or batch release testing. Accordingly, it is normal practice to test at only one temperature, and to assess the results by using an activation energy which is believed to be appropriate for the type of failures which occur. Typically, the value used for mechanisms involving charge accumulation is 0·9 eV, for intermetallic formation it is 0·7 eV and for electromigration it is 0·45 eV.

In order to minimise the duration of the lifetests, and hence reduce costs, it is desirable to use the highest practical stress temperature. The factors which determine the upper limit are still largely a matter of conjecture, but it seems desirable to restrict lifetests to below the glass transition temperature of the moulding compound, because at this point there is a marked increase in both the thermal expansion coefficient of the plastic and the mobility of polar groups within the plastic. For epoxies, the glass transition temperature is typically 150°C and for silicones it is 170°C.

At temperatures not much higher than the glass transition temperature, there is also a risk that failures will be aggravated by thermal degradation of the plastic. This has been demonstrated by a study[5] of how gold/aluminium intermetallic formation is influenced by bromine-based flame retardants and chloride impurities in epoxy moulding compounds. Below 175°C, the bromide and chlorine were chemically bonded to the polymer, but at higher temperatures the polymer–halogen bonds broke, releasing bromine and chlorine to participate in the formation of gold–aluminium intermetallic compounds. Provided the temperature was kept below 175°C the rate of intermetallic-induced failures was acceptable, but at higher temperatures there was a

Table 6.1
Lifetest Durations at 160°C which Simulate
10 Years Service at 90°C

Failure mechanism	Test duration (h)
Charge accumulation	834
Intermetallic growth	2336
Electromigration	8546

drastic increase in the failure rate, and both bromine and chlorine were found to be incorporated into the intermetallic compounds.

The only disadvantage of keeping the stress temperature below the glass transition temperature is that this makes the test duration extremely long for components which are to be operated in equipment where the air temperature is high. The packing density of ICs in many communications and computing/control units is such that the air temperature can reach 70°C and it is not unusual for internal power dissipation within an IC to raise the die temperature to 90°C. For such conditions, the test durations listed in Table 6.1 are needed if a 160°C lifetest is to be used to simulate 10 years in service. These test durations are prohibitively excessive for all but charge accumulation failure mechanisms, so the risk of failure in high temperature applications is usually inadequately assessed. One possible way of rectifying this deficiency is to reduce test times by detecting precursors to failure rather than the failure itself, but this solution has not yet been demonstrated to work.

6.5. THERMOMECHANICALLY INDUCED FAILURES

Thermomechanically induced failures are a consequence of the internal stresses which can be established because of the mismatch between the thermal expansion coefficient of the plastic ($\sim 25 \times 10^{-6}$ per °C), and those of the die and the leadframe ($\sim 12 \times 10^{-6}$ per °C). These stresses build up when the plastic cools down after moulding, and cyclic stresses are induced by changes in temperature during service. Until recently, the most frequently reported failure mechanism caused by the internal stresses was fracture of either the bond wires or the bond between these wires and the IC metallisation. With the advent of LSI dies, however, there have been reports that internal stresses are large enough to cause (i) parametric changes, (ii) damage to both passivation layers and metallisation tracks, and (iii) cracking of the die attach and/or the die. Because of the these problems, the largest area die which can be safely encapsulated using current thermosetting resins is about 7 mm square.

The internal stress due to thermal expansion mismatch has been the subject of detailed investigation in recent years. It has been measured by using on-chip strain gauges[6] and its effects modelled.[7] Particular attention was given to the role of the moulding process and the formulation of the moulding compound; for example, the type and amount of filler have a strong influence on the internal stress. New low-stress resins are being developed but it seems clear that certain compromises have to be accepted; low stress and good thermal conductivity are difficult to achieve in the same formulation, as are low stress and high glass transition temperature.

The only test which can be used to assess the risk of thermomechanically induced failure is arguably the most arbitrary of all those to which plastic encapsulated ICs are subjected. This test involves thermal cycling the components several hundred times between −55 and 150°C with a dwell of 10 min at each temperature. At the end of this stress, outright failures and any thermal intermittent failures are detected by monitoring the operation of the components whilst the temperature is slowly ramped from 0 to 70°C.

Because the thermal cycling test has been used since the introduction of plastic encapsulation, the results of the test provide a 'benchmark' against which to monitor the significant improvements that have subsequently been made to wire bonding control. Products of the late 1960s and early 1970s could only survive a few tens of cycles and bond wire or bond failures in service were a major concern, whereas current products can withstand many thousands of cycles and service failures due to bond wire or bond fracture are rare.

Another form of thermal cycling test, which is gaining popularity for high power dissipation components, is based on varying the die temperature by switching the power supplied to the test specimens on and off with a period which is about twice the thermal time constant of the package ($\sim$7 min). However, it is not yet clear how the results of this thermal cycling test compare with the older test.

The major limitation of the thermal cycling tests is that they do not provide an indication of whether thermally induced stress can cause damage, to passivation, metallisation or die attach, which will subsequently make the component more susceptible to failure mechanisms such as corrosion and electromigration. Thus there is a good case[6] for using special test structures both to optimise the moulding process and to monitor production.

6.6. ALPHA PARTICLE INDUCED FAILURES

Since 1980 there has been increasing concern about the rate at which non-recurrent ('soft') errors can be induced, in memory ICs, by alpha particles emitted by radioactive trace elements in packaging materials. Alpha particles striking the die generate electron-hole pairs which can be separated by the nearby depletion regions of memory elements, and thereby alter the charge which is stored to indicate the logic state. With the increasing levels of integration, the size of the memory elements, and thus the charge stored in them, has decreased making alpha particle induced soft errors a major problem.

Soft error generation by alpha particles was first observed in ICs that were hermetically encapsulated in ceramic packages and these typically produce a flux of between 1 and 100 alpha particles/cm^2 h at the die surface. Proposed remedies include shielding memories with coatings which absorb alpha particles, designing alpha particle immunity into the circuits, and reducing the amount of radioactive materials within the package. The shielding approach has been universally adopted, and it is now common practice to place a thick ($\sim$50 μm) layer of polyimide on the die. Soft error generation rates as low as 100 FITS are now being achieved, whereas in 1980 values of several thousand FITS were being reported.

The fillers used in plastic encapsulants are a source of alpha particles and, at the surface of a plastic encapsulated die, there will typically be a flux of 0·1 alpha particles/cm^2 h. Thus the problem is not as great as that first encountered with some hermetic packages, but nevertheless there is a demand to reduce the levels of radiation in plastic packages (see Chapter 10).

6.7. MOISTURE INDUCED FAILURES

Moisture induced failures have been the most persistent, and thus the most intensively studied, of all the reliability hazards of plastic encapsulated components. The problem arises because plastic encapsulants are unable to prevent ambient moisture from reaching the die.

6.7.1. Ingress of Moisture

There are two methods whereby moisture can reach the die surface, namely diffusion through the bulk of the plastic and ingress along an interfacial leakage path between the leadframe and the plastic. By using a test structure similar to a dual-in-line package, the relative magnitude of these two mechanisms has been assessed[8] for several types of moulding compound, and the results are summarised in Table 6.2.

The final entry in Table 6.2 is for a silicone package which has been subjected to a vacuum impregnation technique in a successful attempt to back-fill (and hence seal) the interface between the plastic and the leadframe. The use of this technique appears to be confined to silicones which, as a class, have larger interfacial leakage paths than epoxies. The large differences between the sizes of the leakage paths for the various types of moulding compound have been ascribed to the degree of chemical bonding of these compounds to the metallic leadframe. Silicone is the only one which does not chemically bond to the leadframe and it thus has the largest leakage paths, whereas epoxy anhydride has the best bonding mechanism and consequently produces the smallest leakage path.

If components are exposed to a chemically hostile environment, the leakage path will be extremely important because it enables ionic species to reach the die. Users of the increasingly popular surface mounted packages are likely to have to pay particular attention to this hazard. For these packages, the leadframe is shorter than in the

Table 6.2
Comparison of Bulk and Interfacial Ingress
Rates (at 20°C/100% RH)

Plastic	Permeation rate (μg/week)	
	Bulk	Interfacial
Epoxy anhydride	45	21
Epoxy novolac	250	100
Silicone	425	650
Back-filled silicone	450	0

standard dual-in-line format, and the surface mounting process often involves cleaning in chlorinated solvents in order to remove flux from underneath the package.

Even if the leakage path is totally eliminated by back-filling, moisture is still able to reach the semiconductor by permeation through the plastic itself. The two aspects of bulk permeation which have to be considered are (i) the amount of moisture which a plastic can absorb from the ambient, and (ii) the time taken for a significant concentration of moisture to build up at the die surface.

The equilibrium level of water absorption has been assessed[8] by monitoring the weight gain of 1 mm thick moulded discs exposed to various combinations of temperature and relative humidity (RH). To a first approximation the increase in weight was found to be independent of temperature and proportional to the square of the RH, over the ranges 23–100°C and 40–90%, respectively. As a group, epoxies absorb considerably more water than silicones; for example, an 80% RH will cause epoxies to increase in weight by about 0·5%, whereas the corresponding figure for silicone is about 0·1%.

The absorption of water by the plastic causes two changes which have a direct bearing on reliability: it increases the mobility of ionic impurities and, in making the plastic expand, it increases the likelihood of the formation of a gap between the plastic and the semiconductor die. The importance of these two factors will be discussed in the next section.

The time dependence of the absorption of water has also been studied using the weight gain studies,[8] and it is apparent that it can be described by a diffusion constant, D, which is independent of RH. The average results for four epoxy novolac and anhydride moulding compounds can be summarised by the equation

$$D = 5·6 \times 10^{-3} \exp\left(\frac{-3800}{T}\right) \tag{6.6}$$

where T is the temperature (K). For the silicones studied, the rate of ingress was too rapid to permit accurate measurements to be made.

By using eqn (6.6) it is possible to show that the time taken for the moisture level at the die surface to approach its equilibrium concentration is short compared with the period that an IC is expected to operate in service. For example, with an epoxy dual-in-line package at 20°C, it will only take 300 h for the concentration of water at the die to reach 90% of its equilibrium value, and the corresponding figure for 85°C is

40 h. It follows then that the time taken for moisture to reach the semiconductor die is not the rate-limiting factor for moisture induced failures.

6.7.2. Effect of Moisture on an Unpassivated Die

The effect of moisture is best illustrated by first considering the results obtained for a test specimen which has a metallisation pattern that is not protected (i.e., covered) with a passivation layer. This structure has the advantage of providing a means of directly observing the changes which are a precursor to failure, and it is also representative of most discrete components.

By using unpassivated interdigitated aluminium tracks deposited on a silicon dioxide layer, the surface conductivity of plastic encapsulated dies has been monitored[9] during accelerated lifetests based on a combination of high temperatures and high relative humidities. Figure 6.4 shows a simplified plan view of the test structure, and a set of results obtained for an epoxy novolac package which was exposed to 110°C/90% RH. It can be seen that after about 30 h the surface conductivity started to rise, thereby indicating that moisture had arrived at the die, and that it continued to increase for a further 30 h in response to the accumulation of water on the die surface. The eventual

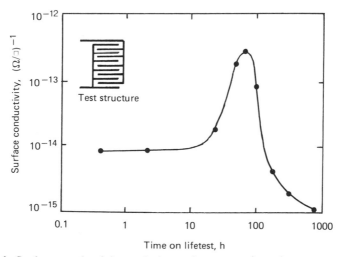

Fig. 6.4. Surface conductivity variations of an unpassivated test structure (see insert) during a 110°C/90% RH lifetest.

decrease in surface conductivity was attributed to the formation of insulating oxides on the metal tracks and/or ionic space charges which reduced the fields at the tracks.

The mechanism by which a conducting film of water forms on the surface of the test dies has been investigated[9] by comparing the behaviour of the surface conductivity of interdigitated structures encapsulated in a variety of moulding compounds. It has been shown that a conducting film of water forms more rapidly with epoxy novolacs than it does with epoxy anhydrides, and that it is seldom observed with silicones or silicone–epoxies. Moreover, the relationship between surface conductivity and RH was found to be similar to that measured when an unencapsulated die was exposed to moisture. From these observations it is concluded that the conducting film only forms if there is a gap between the plastic and the die, and that the RH in the gap can approach that of the ambient of the package. The existence of a gap between the plastic and the die after exposure to humidity stress tests has been confirmed by cross-sectioning and by on-chip strain gauges, but it is still a matter of speculation as to whether the gap forms due to swelling of the plastic and/or the breaking of adhesive bonds between the plastic and the die. There is also no consensus on whether the formation of the gap should be referred to as 'debonding' or 'delamination'!

The current flowing in the water film that links the metal tracks of the test structure causes electrochemical corrosion which eventually results in the tracks going open circuit. The magnitude of this current, and hence the time to failure, is strongly influenced by the amount of ionic impurities leached out of the plastic. These ionic impurities can also form space charges that produce accumulation or depletion of the underlying silicon, and thereby cause active elements to fail due to grain degradation, reduced breakdown voltage or channelling.

Ionic impurities leached out of the plastic have been shown to make the rate of degradation more rapid than that observed for unencapsulated dies exposed to the same conditions. In addition, by breaking down the protective oxides which form on aluminium, ionic impurities can change the corrosion from being predominantly cathodic to being predominantly anodic, and ionic impurities which are gold complexing agents (e.g., chlorine) can cause corrosion of gold metallisation tracks.

The influence of the ionic contamination leached from the plastic can be illustrated by 85°C/85% RH lifetest results for unpassivated transistors which were encapsulated in several types of moulding

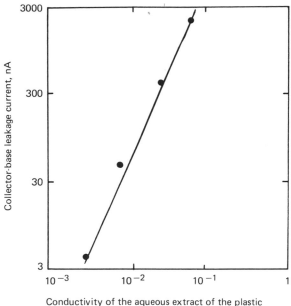

Fig. 6.5. Relationship between the collector-base leakage current of transistors subjected to a 1000 h 85°C/85% RH lifetest, and the conductivity of the aqueous extract of the plastic encapsulant.

compound.[10] Figure 6.5 shows that the collector-base leakage current, measured after 1000 h at 85°C/85% RH, was correlated with the ionic content of the moulding compound as determined by an aqueous extract conductivity test method. (The latter, as its name suggests, is based on measuring the conductivity of water drawn off from a cell in which a mixture of water and plastic are held at 120°C for 24 h.) The aqueous-extract conductivity test has been used to assess the trend in ionic contents over the last decade.[11] The results, given in Fig. 6.6, show that the ionic impurity content of silicones has been consistently low but has now been bettered by novolacs, and that anhydrides have always been the least pure.

Finally, having placed particular emphasis on the hazard posed by ionic impurities in the plastic, it is only fair to stress that the encapsulated die is also at risk from ionic species which can gain entry from the exterior of the package via leakage paths along the leadframe, and

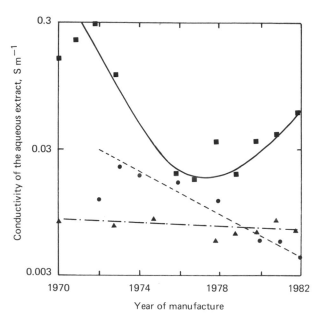

Fig. 6.6. Historical trends in the conductivity of the aqueous extract of: ■, epoxy anhydride; ●, epoxy novolac; and ▲, silicone moulding compounds.

from ionic contamination which may be present on the die surface prior to encapsulation. Thus, in addition to taking care over the purity of the materials used, it is essential that attention be given to assembly hygiene and the chemical hazards experienced by the completed package. If ionic contamination can reach the die from any of these sources, the only safeguard against ionically enhanced moisture induced failure is the lack of a gap between the plastic and the die. Should such a gap exist after encapsulation, or if delamination occurs during service, the only method of protecting against moisture induced failures is to cover the die with a passivation layer.

6.7.3. Protection Provided by Passivation

In the ideal case of a passivation that is free of pin-holes and cracks, a film of water which forms on the surface of the die will only come into contact with the metallisation layer at the windows cut to permit bonding. Fortunately, this is the region where there is least risk of failure due to electrochemical corrosion, firstly because the field strengths are lower than for the fine metal tracks over the rest of the die and secondly because the volume of aluminium is greater than for such tracks. However, if the passivation contains cracks or pin-holes it is possible for the water film to provide a conducting path between tracks, and corrosion failures will occur unacceptably early in the life of the component.

The first passivation layers were pure silicon dioxide, produced by a chemical vapour deposition (CVD) process. Later phosphorus was added, in uncontrolled amounts, to relieve strain in the layer and thus prevent cracking and loss of adhesion. By 1974 it had been established that, in order to minimise aluminium corrosion in a moist environment, the optimum amount of phosphorus in a silicon dioxide layer is about 2% by weight. The relationship between phosphorus content and time to failure is shown by the typical set of results[9] presented in Fig. 6.7. Below the 2% level, the time to failure is an increasing function of the amount of phosphorus, because the latter reduces cracking and improves adhesion. Above the 2% level, the time to failure decreases

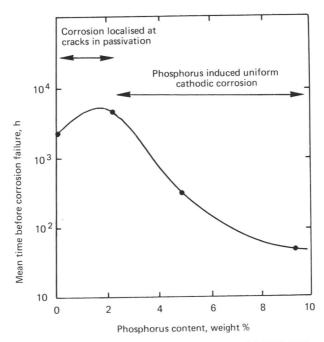

Fig. 6.7. Median life for aluminium corrosion, at 110°C/90% RH, as a function of the phosphorus content of silicon dioxide passivation.

because the phosphorus takes an active role in the corrosion mechanism in one of two ways:

(1) Phosphorus migrates to the surface of the passivation and, if a water film is able to form on the surface of the die, it enhances the rate of bond-pad corrosion and of corrosion associated with pin-holes or cracks in the passivation. Such corrosion is highly localised and can affect anodic and/or cathodic tracks, depending on the type and amount of other forms of ionic contamination which may be present.

(2) Water which permeates the passivation makes the latter conducting, and thus it is no longer necessary to have a film of water on the surface of the passivation in order for corrosion currents to flow. Corrosion produced by this second mechanism is easy to identify because it usually affects all cathodic tracks. In the early stage, only the high field points (i.e., ends of tracks) are attacked, but the corrosion rapidly becomes uniform over all such tracks.

The hazards associated with phosphorus-doped silicon dioxide passivations are now widely recognised and manufacturers have accordingly adopted control limits of 2–4% by weight of phosphorus. A few manufacturers are also making use of alternative forms of passivation layers, the three most popular of which are:

(1) Undoped CVD silicon dioxide which is thin enough (2000 Å) to withstand the built-in stress without cracking.
(2) Silicon nitride produced by plasma-assisted CVD.
(3) Polyimide.

Silicon nitride has received the most attention, but only a few manufacturers are using it commercially because it is difficult to control some of the properties which can affect the operation of an IC (such as hydrogen content and stress). The main attribute of silicon nitride is that it is impervious to moisture and this makes it the ideal passivation when highly phosphorus-doped silicon dioxide layers are used to separate two levels of interconnection, as, for example, in gate arrays. (One requirement of an inter-layer dielectric is that it should smooth out underlying steps, and 6–10% phosphorus-doped silicon dioxide has the appropriate flow properties to do this.) In such a case silicon nitride prevents the uniform cathodic corrosion which is associated with the use of such oxide layers as passivations.

By using the appropriate, defect-free passivation in conjunction with

Table 6.3
Conditions Which Will Result in Unacceptable Moisture Induced Failure Rates

Plastic	Passivation	Failure mode
Poor adhesion to die		Bond-pad corrosion
High ionic content	Cracks and/or pin-holes	Localised corrosion
	High phosphorus content	Uniform cathodic corrosion
Or any combination of the above		

a plastic which adheres well (and permanently) to the die, moisture induced degradation can be prevented. However, it is possible to accept inferior performance provided the failure rate is tolerable and this is the situation which will usually be encountered.

Because the defects of some plastics can be partially offset by good passivations, and vice versa, there will be various combinations of moulding compounds and passivations which, although allowing some degradation to take place, nevertheless offer adequate reliability in moist ambients. For example, although bond-pad corrosion will occur if there is a loss of adhesion between a plastic and a defect-free passivation, the failure rate will only be unacceptable if there is an abundance of ionic contamination. The combinations which should be avoided are those where the rate of degradation will be intolerably high—these combinations are detailed in Table 6.3.

Although the last decade has seen sustained improvements, in both moulding compounds and passivation layers, which have lessened the risk of moisture induced failure, it is still considered necessary to perform regular product evaluation.

6.7.4. Standard Lifetest
A 1000 h 85°C/85% RH lifetest has become the standard industry method of assessing the moisture resistance of plastic encapsulated components. Although the test conditions were initially chosen somewhat arbitrarily, they can be justified by using empirical equations derived from studies which have shown that the rate of moisture induced degradation is a function of both relative humidity and temperature. All but one of these equations are derived by fitting curves to

experimental data, whereas the exception[12] is based on the hypotheses that: (i) the surface conductivity of the die is the rate-limiting factor, and (ii) the dependence of this factor on both relative humidity and temperature will be the same for both a plastic encapsulated die and an unencapsulated (bare) die. The latter premise is a reasonable assumption for the case where a gap forms between the moulding compound and the die, but there are two reasons for arguing that the theoretical basis for the model is little better than for the curve fitting approaches. Firstly, the equations used to model the surface conductivity of the bare die are empirical. Secondly, there will be some cases where failure can occur without there being a gap between the plastic and the die, as when the passivation has a high phosphorus content or when there are cracks in the passivation.

The various equations used to describe moisture induced degradation differ considerably in form, but they all predict similar behaviour over the range of high relative humidities used to study moisture induced degradation (i.e. 70–95%). One of the simplest of these equations, and the one which arguably gives the best fit to the available experimental data, is based upon a $(RH)^2$ term to describe the relative humidity dependence, and upon a simple Arrhenius expression for the temperature dependence. This equation[10] can be written as

$$\text{Time to failure} = \exp\left(-Ah^2 + \frac{\Phi}{kT}\right) \qquad (6.7)$$

where h is the relative humidity expressed as a fraction, T is the temperature (K), k is Boltzmann's constant and A and Φ are data-fitting parameters. The value of A which gives the best fit to the published data is 4·4. Depending on the type of failure, the value of the activation energy term Φ can vary[13] from 0·6 to 2 eV, but a value of 0·6 eV is usually used when assessing plastic encapsulated components because it has been found to be characteristic of products having the lowest moisture resistance. By substituting these values into eqn (6.7) it is possible to derive the acceleration parameter A which will relate the time to failure in operation conditions (T_0 K and a RH of h_0) to that for lifetest conditions (T_s K and h_s)

$$\text{Acceleration factor} = \exp\left[4\cdot4(h_s^2 - h_0^2) - 6960\left(\frac{1}{T_s} - \frac{1}{T_0}\right)\right] \qquad (6.8)$$

The voltage applied to the components during the lifetest will also influence the time to failure, but the published data are scarce. Although it has been established that the moisture induced failure rate increases with the magnitude of the intertrack electric field, no agreed relationship has emerged. Accordingly it is now usual practice to bias the lifetest specimens as they would be in service.

The empirical equations are used extensively to relate performance on lifetests to that which would be obtained under the more benign service conditions. However, it should always be appreciated that the experimental data on which to judge their accuracy are limited. The $(RH)^2$ equation has been shown to be in reasonable accord with the relative responses to tests performed at 85°C/85% RH and the 'tropical climate' conditions of 27°C/90% RH but, as with the other equations, there is a need to extend the comparison to the lower relative humidities encountered in temperate climates. Unfortunately, such studies will be difficult to perform, because the improvements in plastic encapsulation have made the test times needed to make such comparisons extremely long.

Although precision cannot be claimed for calculations based on the empirical equations, these provide the only means of predicting the influences of low relative humidities and temperatures, and the guidelines which result are invaluable. One of the most important of these predictions is that power dissipated in the die will significantly increase the time to failure corresponding to a given amount of water in the ambient. This improvement is brought about because self-heating raises the temperature of the die and thereby lowers the effective relative humidity at the die surface. In order to make allowance for the influence of this self-heating, it is sufficient to ensure that the value of RH which appears in eqn (6.8) is that which would be measured at the die surface. To calculate this RH, it is necessary to know both the dew point of the ambient and the die temperature, and these two parameters form the basis of a simple way of graphically expressing the degree of assurance gained by performing a given lifetest.

By using eqn (6.8) it can be shown that a 1000 h 85°C/85% RH lifetest will simulate 20 years service at the operating conditions defined by the continuous line on Fig. 6.8. The axes are the dew point of the ambient and the temperature of the die. For conditions which lie above the line corresponding to the required service life the humidity stress test is too short, whereas for conditions below the line the test is unnecessarily severe.

The broken line on Fig. 6.8 defines a region within which eqn (6.8) is

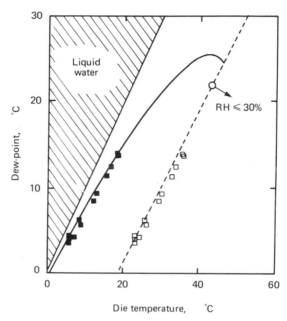

Fig. 6.8. Operational conditions for which 20 years in service is equivalent to a 1000 h lifetest at 85°C/85% RH. (For a shorter service life the duration of the lifetest can be reduced on a pro rata basis.) The data points indicate the monthly average conditions for a typical UK location for: ■, an unheated die, and □, a die whose temperature is raised 16°C by self-heating.

invalid. To the right of the broken line, the RH at the die surface is less than 30%, and for components operating within this region it is believed that the failure rate due to moisture induced degradation will be too small to be of any practical consequence.

In order to relate Fig. 6.8 to the conditions likely to be encountered in service, the shaded portion is added to define the region corresponding to the (hopefully) unlikely case of operation in liquid water, and the data points indicate the monthly average conditions for a typical UK uncontrolled environment (i) when there is no internal heating, and (ii) when such heating is just sufficient to raise the temperature of the die by an amount (16°C) to ensure that the relative humidity at the die surface will always be less than 30%. From this data it can be concluded that a 1000 h 85°C/85% RH lifetest will simulate 20 years operation in any UK environment. However, by performing a similar exercise for tropical climates (typically 27°C/90% RH) it can be shown that the duration of the lifetest would have to be raised to 7000 h in order to simulate 20 years in such conditions.

From Fig. 6.8 it also follows that there will be no risk of moisture induced failure occurring in service in the UK, if internal power dissipation raises the die temperature by only 16°C above the uncontrolled ambient temperature. (The corresponding figure for tropical climates is 18°C.) Although there are many technologies (such as bipolar TTL) which satisfy this criterion, there are others (such as CMOS) where the internal heating is minimal, and consequently ambient moisture will present a significant reliability hazard.

Between these two extremes there are cases where the amount of internal heating is sufficient to make it practical to reduce the duration of the lifetest and yet still achieve the required degree of assurance in predicting service performance. However, because the 1000 h 85°C/85% RH lifetest has become the 'bench-mark' against which improvements in plastic encapsulation have been judged, it can be argued that reducing the duration of the test for some specimens will result in the loss of the practical and historical (i.e., traceability) advantages of a standard test.

6.7.5. Results from the Standard Lifetest

Many, if not all, manufacturers have the technology needed for the production of plastic encapsulated ICs which are capable of reliable operation in moist environments, but the timing of its introduction is governed both by economics and by the collective pressures of reliability conscious users. This can be illustrated by the results obtained for groups of CMOS ICs purchased from one manufacturer during the period 1981 to 1983. Until recently, the percentage failures noted after the product had been subjected to the 1000 h 85°C/85% RH lifetest varied from 5 to 64%, with the average being 30%. However, when the manufacturer decided to supply the professional equipment market, improvements in passivation were made and the product is now consistently able to withstand the standard lifetest without producing any failures. It is, perhaps, also worth noting that had the worst product been used in an uncontrolled UK ambient the failure rate after 10 years in service would have been 500 FITs; a value which is in marked contrast to the acceptable level of about 20 FITs for such components.

From other lifetests on a wide variety of ICs, in both the traditional dual-in-line format and the emerging range of low-cost surface-mounted packages, it is clear that poor performance is usually the result of insufficient attention being paid to the passivation. Many manufacturers recognise the need to monitor this aspect of the process and, accordingly, regularly perform lifetests on their high volume product. Purchasers from less conscientious manufacturers run the risk of obtaining an unreliable product unless they impose (and usually pay for) lot-release testing.

Although manufacturers can now produce plastic encapsulated components which are likely to have satisfactory reliability in temperate climates, considerably more effort is needed to produce components which are suitable for tropical climates (i.e. 30°C/90% RH). Thus there is likely to be a very limited availability of plastic encapsulated components suitable for such an application.

6.7.6. Rapid Assessment Tests
The 1000 h 85°C/85% RH lifetest is generally agreed to be the standard method of assessing the risk of moisture induced failure. However, because it is unsatisfactory to have to wait 1000 h (6 weeks) for a result, there is some opposition to using this lifetest for either process control or batch release testing. Accordingly, there has been considerable interest in reducing the test duration by using more severe stress conditions. Several test methods have been developed, and their common feature is that the partial pressure of water in the test chamber is greater than 1 atm. Thus it is necessary to use pressure vessels (autoclaves) for the 'rapid' test methods, instead of the environmental chambers used for the standard 85°C/85% RH lifetest.

The rapid test methods either involve unbiased storage in saturated conditions (up to 131°C/100% RH) or are based upon lifetests in very high temperatures and relative humidities (up to 150°C/81% RH). The first of these options has long been favoured by semiconductor manufacturers because the required conditions can be produced easily and there is no need to apply bias; an autoclave containing water and the test specimens is simply placed into a constant temperature oven. The second option has been a research technique for many years but is now finding more widespread use thanks to the recent availability of commercially made test chambers; these are autoclaves in which it is possible to establish two temperature zones and thereby produce an unsaturated condition.

In order to emphasise the advantages and limitations of the rapid test methods, a comparison will be made of the results obtained when these techniques, and a standard 1000 h 85°C/85% RH lifetest, were performed on CMOS quad two-input NAND gates from six manufacturers who will be designated A to F. In all cases, the moulding compound was an epoxy novolac; one of the manufacturers (F) used a nitride passivation and the rest phosphorus-doped oxide.

For each manufacturer, 50 specimens were subjected to a 1000 h 85°C/85% RH lifetest and further samples of size 15 were subjected to 90% RH lifetests, and to saturated storage tests at temperatures within the range 110–131°C. In addition, the events which led to failure were monitored by using a capacitance technique to measure the surface conductivity of the die and thus to detect delamination between the plastic and the die. The results of this study[13] are summarised in Table 6.4, which lists the failure fractions after a 1000 h 85°C/85% RH lifetest, a 100 h 131°C/90% RH lifetest and a 350 h saturated storage test. Each of these conditions caused some specimens to delaminate, but it was only in the case of the saturated storage test that delamination was a necessary precondition for failure. The average surface conductivity of the specimens which delaminated during the saturated storage tests is listed in the final column of Table 6.4.

From this data it is clear that the 85 and 131°C lifetests both ranked the manufacturers in the same order, and that the results of these tests were totally at variance with those of the saturated storage tests. The explanation for this behaviour was obtained from the delamination data and from an analysis of the type of corrosion which caused the failures on the various tests.

Table 6.4
Summary of High Temperature/Humidity Tests on CMOS ICs from Six Manufacturers

Manufacturer	*Lifetests* *Failure fraction after*		*350 h Saturated* *storage at* 131°C	
	1000 h 85/85	*100 h* 131/90	*Failure* *fraction*	*Surface* *conductivity* $(\Omega/square)^{-1}$
A	32/50	15/15	1/15	1.3×10^{-11}
B	4/50	9/15	0/15	3×10^{-11}
C	1/50	3/15	0/15	7×10^{-12}
D	0/50	0/15	1/15	4×10^{-11}
E	0/50	1/15	8/15	$>10^{-10}$
F	0/50	3/15	9/15	$>10^{-10}$

Failures during the saturated storage tests were invariably preceded by delamination, and were always caused by corrosion of the exposed aluminium at the bond-pads. This type of corrosion during saturated storage tests is attributed to galvanically generated currents flowing in a layer of water linking the gold ball bonds and the surrounding unpassivated aluminium. It is not surprising, therefore, that the data presented in the last two columns of Table 6.4 show that the failure fraction was strongly dependent on the surface conductivity of the die after delamination. Thus the rate of failure during saturated storage tests is determined by the amount of ionic impurities which can accumulate in the water film that forms on the die surface after delamination, and is primarily dependent upon the ionic impurity content of the plastic.

The relationship between delamination and failure for the lifetested specimens was more complex, and it was only for manufacturer F (the one that used a nitride passivation) that (i) delamination was a necessary precondition for failure, and (ii) the corrosion was restricted to the exposed aluminium at the bond-pads. For the other manufacturers, less than 50% of the lifetest failures delaminated prior to failure and, although the delaminated specimens showed signs of bond-pad corrosion, there were two other types of corrosion which were more prevalent. On the early failures there were small areas of severe corrosion scattered over the die and the frequency of such sites was correlated with that of passivation crack and pin-hole density for each manufacturer (up to 36 per die in the worst case), whereas for the late failures (and survivors) there was uniform corrosion of all the cathodically biased tracks. This uniform corrosion was similar to that usually ascribed to high levels of phosphorus in CVD silicon dioxide passivations, so it would appear that this type of corrosion can even occur when, as in the cases reported here, the phosphorus contents are within the range 3·1–5·5%.

The difference between the results of the saturated storage tests and the lifetests can thus be explained as follows. Failures during the saturated storage tests only occur after delamination between the plastic and the die, and are related to the impurity concentration of the plastic, whereas failures during lifetests will usually be attributed to passivation defects and in this case delamination is not a necessary precondition for failure. Thus saturated storage tests only assess the plastic, while lifetests assess both the plastic and the passivation. With some combinations of plastic and passivation, the two types of test will give similar ranking of quality, but this will not be the case for all combinations.

Despite the reservations about the scope of the saturated storage test, the latter is very popular with semiconductor manufacturers because it is a quick and simple method of monitoring some of the factors which will influence the reliability of components operating in moist environments. However, there are indications that the semiconductor manufacturers are also beginning to make more use of the rapid lifetests, although there is as yet no consensus on what the test conditions and duration should be.

Because the rapid lifetests are usually based on relative humidities within the range 81–90%, most of the acceleration relative to the standard 85°C/85% RH lifetest is achieved by using test temperatures which are substantially in excess of 85°C. In order to estimate the amount of acceleration, and thereby fix the test duration which is equivalent to 1000 h at 85°C/85% RH, it is necessary to know the value of the activation energy describing the moisture induced degradation, and unfortunately this can vary from manufacturer to manufacturer. In the study[13] upon which Table 6.4 was based, a value of 0·6 eV was recorded for the specimens which were the most susceptible to moisture (manufacturer A); for those with the best performance (manufacturer D) it was 2·2 eV and for the others it ranged from 0·8 to 1·0 eV. The simplest solution to this dilemma is to base the duration of the rapid lifetests on the 0·6 eV activation energy which appears to be characteristic of the weakest product. On this basis, for example, a 200 h lifetest at 110°C/85% RH would be equivalent to the standard 1000 h lifetest at 85°C/85% RH. Shorter duration tests can be achieved by making the temperature greater than 110°C but, because the product which is the least susceptible to moisture has the highest activation energy,[13] the difference between the failure data for various manufacturers becomes less pronounced as the test temperature is raised. Thus there is always a risk that 'rapid' lifetests may be less discriminating than the standard 85°C/85% RH lifetest, if the test temperature is made too high.

6.8. PRESENT STATUS

Whilst it is generally accepted that the present day plastic encapsulated components are considerably more reliable than those made several years ago, this confidence is based largely on evidence which has been generated by stress testing rather than by service experience. Nevertheless, the use of such components by the computer and tele-communications industries (i.e., the non-military, professional sector) has increased dramatically during the last two years, and it remains to be seen whether the initial advantage of low cost (relative to hermetically encapsulated devices) has been gained without having to sacrifice service performance. There have been unsubstantiated reports that some professional users have already experienced embarrassingly high failure rates with plastic encapsulated components, but these problems appear to involve the use of unassessed products in high humidity environments (i.e., tropical or semi-tropical climates).

It is now generally recognised that the key elements of plastic encapsulation technology are:

(1) a passivation layer that adheres well to the die surface is struc-turally sound (i.e., is free from cracks and pin holes), and which does not contain 'active' ionic impurities

(2) high and consistent wire bond strengths (normally achieved using fully automated wire bonders)

(3) a moulding compound which adheres well to the die surface, and has a low ionic content, water adsorption and thermal expansion and a high glass transition temperature

Unfortunately, the priorities of the semiconductor manufacturers and of reliability-conscious users are somewhat different. Semiconduc-tor manufacturers have made improvements in bonding in order to increase yield, but their only other principal requirement is ease of moulding of the plastic encapsulant. Thus changes which improve reliability are not usually made until the reliability-conscious users become a significant fraction of the market for a given technology.

At present, there are few reliability reservations about bipolar TTL and LPSTTL; these are made essentially for the professional market and semiconductor manufacturers have responded to the pressure for a better product. Power transistors in general are also satisfactory, they use a unimetallisation/wire bond system (Au/Au or Al/Al) and the encapsulant is normally a silicone. Small signal transistors are largely aimed at the non-professional market; their quality and reliability are usually poor (but adequate for their intended application). CMOS and ISOCMOS are particularly sensitive to moisture, but some manufac-turers have demonstrated that they can produce a reliable product and, as these technologies are widely used in professional applications, it is almost inevitable that other manufacturers will follow this lead. However, it must be recognised that very severe lifetests are required to show that such components are suitable for operation in tropical climates. NMOS is now almost exclusively used for LSI and VLSI circuits, and as these are generally high power dissipation components, the major reliability concerns are thermally induced and stress induced failures.

The most persistent reliability problem with all plastic encapsulated components has been the wide variation in quality and reliability of products from different manufacturers and between sample lots from the same manufacturer. With improvements in technology, this variation is likely to become less, but if a purchaser is not discriminating he will always run the risk of obtaining a product which has been rejected by a more discerning customer. Reliability conscious users purchase from approved vendors who have demonstrated that their technology is right, and who perform quality monitoring and batch release tests which are likely to give an indication of service performance. Although approval exercises and regular testing are an added complication, it is nevertheless possible to derive test schemes which do not negate the principal attraction of plastic encapsulated components, namely their low cost.

ACKNOWLEDGEMENTS

I gratefully acknowledge the support and guidance of R. W. Lawson and F. H. Reynolds. Acknowledgement is also made to the Director of the British Telecom Research Laboratories for permission to publish.

REFERENCES

1. Taylor, C. H., Plastic encapsulated semiconductor devices—bibliography III, *Microelectron. Reliab.*, **19** (1979) 403.
2. Schnable, G. L. and Comizzoli, R. B., CMOS integrated circuit reliability, *Microelectron. Reliab.*, **21** (1981) 33.
3. Berg, H. M. and Paulson, W. M., Chip corrosion in plastic packages, *Microelectron. Reliab.*, **20** (1980) 247.
4. Reynolds, F. H., Thermally accelerated aging of semiconductor components, *Proc. IEEE*, **62** (1974) 149.

5. Gale, R. J., Epoxy degradation induced Au–Al intermetallic void formation in plastic encapsulated MOS memories, *Reliability Physics Symposium Proceedings*, **22** (1984) 37.
6. Schroen, W. H., Spencer, J. L., Bryan, J. A., Cleveland, R. D., Metzgar, T. D. and Edwards, D. R., Reliability tests and stress in plastic encapsulated circuits, *Reliability Physics Symposium Proceedings*, **19** (1981) 81.
7. Ubbing, J., Mechanisms of temperature cycle failure in encapsulated optoelectronic devices, *Reliability Physics Symposium Proceedings*, **19** (1981) 149.
8. Harrison, J. C., Control of the encapsulation material as an aid to long term reliability in plastic encapsulated semiconductor components, *Microelectron. Reliab.*, **16** (1977) 233.
9. Sim, S. P. and Lawson, R. W., The influence of plastic encapsulants and passivation layers on the corrosion of thin aluminium films subjected to humidity stress, *Reliability Physics Symposium Proceedings*, **17** (1979) 103.
10. Lawson, R. W., The accelerated testing of plastic encapsulated semiconductor components, *Reliability Physics Symposium Proceedings*, **12** (1974) 243–9.
11. Lawson, R. W., A review of the status of plastic encapsulated semiconductor component reliability, *British Telecom Technol. J.*, **2** (1984) 95.
12. Sbar, N. L. and Kozakiewicz, R. P., New acceleration factors for temperature, humidity, bias testing, *IEEE Trans. Electron Devices*, **ED-26** (1979) 56.
13. Merrett, R. P., Bryant, J. P. and Studd, R., An appraisal of high temperature humidity stress tests for assessing plastic encapsulated semiconductor components, *Reliability Physics Symposium Proceedings*, **21** (1983) 73.

Chapter 7

POLYMERS FOR SEMICONDUCTOR PROCESSING

S. CLEMENTS*

Plessey Research (Caswell) Ltd, Caswell, Towcester, UK

7.1. INTRODUCTION

In many applications of polymers their susceptibilities to degradation or embrittlement by radiation and attack by some solvents are considered shortcomings. In lithographic applications, however, these properties are used to advantage. The processing of semiconductors is achieved entirely by the selective degradation or crosslinking of polymeric coatings and each step of etching, doping, passivation or interconnection of the substrate is carried out using this imaged protective coating or resist. The patterning of the resist is usually conducted by a photographic process using near ultra-violet light, followed by selective dissolution of either exposed or unexposed areas.

The principle involved is far from new because this same technique was used in the 1820s by Niepce to produce what is regarded as the first photograph. Using a solution of asphalt in white spirit he coated glass plates with a thin layer of asphalt by evaporation of the spirit. When this was given an extended exposure to light, the ultra-violet component of the radiation hardened the film and reduced its solubility to organic solvents. Careful treatment in a suitable solvent, in this case oil of lavender, removed unexposed and thus unhardened asphalt leaving the insoluble exposed regions. Niepce also found that if he coated a metal plate it could be etched with acid where unprotected by the primitive resist and used for printing. The process, known as

* Present address: FTL, London Road, Harlow, Essex CM17 9NA, UK.

heliogravure, was used for copying prints but was very slow. It is interesting to note, however, that modern ultra-violet resist materials are processed in virtually the same manner: coated from solution, exposed to radiation and developed in a solvent.

The polymers currently used were originally developed for the printing industry and have changed little in principle in 30 years.[1] Naturally the requirements of the resist materials in microlithography, where patterns are defined with dimensions of a few microns, are much more stringent than those in the printing industry where the dimensions are commonly two orders of magnitude larger. Hence the resists have been refined to some extent to achieve the required resolution but the sensitivity, vastly improved over Niepce's asphalt, is quite sufficient for processing semiconductors. This is quite fortunate because sensitivity is a primary factor in determining throughput in microlithography whereas in the printing industry the resist is used only to produce a master plate from which many thousands of prints are made and sensitivity is not crucial.

In semiconductor manufacture the resist is spun from solution at several thousand revolutions per minute to form a film of uniform thickness between 0.5 and 2.0 μm and exposed to one of a variety of forms of radiation including X-rays and ion or electron-beams as well as near or deep ultra-violet. Selective irradiation is achieved by either exposing through a mask or writing with a focused 'pencil' beam. The resist materials are then classed as either negative or positive acting depending on their response to radiation. A resist is negative acting if, as in Niepce's material, exposed areas are rendered less soluble, usually by crosslinking, and positive acting if irradiated areas become more soluble, usually because of some degrading reaction (Fig. 7.1). The substrate is subsequently etched, doped or deposited with metal as necessary before finally removing or stripping the resist by wet or plasma methods. Usually these steps are interspersed with baking steps to remove stress in the polymer before exposure, to promote substrate adhesion or to improve resistance during processing.

The trend in the electronics industry is to ever smaller geometries in order to improve device operating speeds and this demands not only higher resolution but also the need to withstand highly corrosive reagents and, increasingly, the rigours imposed by dry processing. Fortunately polymers, or polymer/sensitiser combinations, are well suited to most of these requirements including film-forming properties of homogeneity, planarisation, adhesion, etc., which are often over-

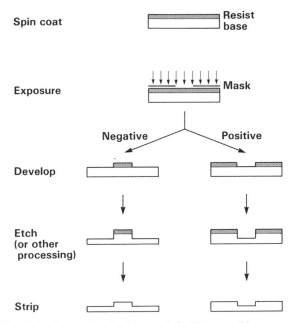

Fig. 7.1. Generalised scheme of the lithographic process.

looked when considering resist materials. These properties and the chemistry involved will become evident as the lithographic process is studied in more detail.

7.2. RESIST REQUIREMENTS AND POLYMER PROPERTIES

Resist materials vary widely in their properties according to the requirements of the lithographic process which may employ different exposure sources, different minimum feature sizes and a range of processing conditions including wet and dry etching media and baking steps. These requirements are classified in a few primary characteristics common to all resists which broadly describe any particular material and an understanding of the meaning and measurement of these fundamental properties is essential before the various materials can be described in detail. Many of the film-forming or secondary properties of resists are a consequence of the nature of polymers and it will be seen that the primary resist characteristics are also influenced by polymer parameters and in particular their physical and solution properties.

7.2.1. Sensitivity and Contrast

In simple terms the lithographic sensitivity of a resist is a measure of the exposure required to delineate the resist pattern. In practical terms, however, the actual exposure required depends on a wide number of factors so that figures quoted for the sensitivity of the same resist tend to differ between users, not least because they are defined in different ways. Naturally the sensitivity is primarily determined by the concentration and efficiency of the radiation-sensitive groups in the resist film be they sensitive to photons, electrons or otherwise and the nature of these groups will be described later for each lithographic system. In addition, however, the required exposure dose is affected by polymer parameters such as molecular weight, process details and the development process in particular.

7.2.1.1. Negative Resists

When a conventional negative tone resist is exposed, crosslinking of the polymer matrix is induced, increasing the molecular weight. As a result of the increased bonding, the solubility in a given solvent decreases and in order to develop out an image the crosslink density must be sufficiently high that it is rendered insoluble. It can be immediately seen that the choice of developer solvent is fundamental in determining whether an image is produced at all and also exactly what dose is required. At low doses the crosslink density is low enough for the solvent to still dissolve and hence remove the resist. At some threshold dose known as the gel dose D_n^i (Fig. 7.2) the solvent can penetrate the polymer matrix but not solubilise it. Doses above this value lead to an insoluble residue, which remains after development if the adhesion to the surface is sufficiently strong. The gel dose is strongly dependent on molecular weight and an order increase in the molecular weight results in a decrease of an order in the gel dose and consequent sensitivity.[2,3] The practical limits to the molecular weight are set by resolution considerations discussed later and the ability to dissolve the polymer to give solutions which are suitable for spinning and development.

Sensitivity is conveniently measured by exposing resist films to a range of doses and measuring the thickness of resist remaining

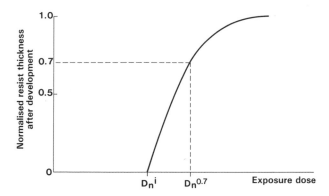

Fig. 7.2. The effect of exposure dose on the developed resist thickness for a negative resist.

(Fig. 7.2). In practice, exposure to doses sufficient to result in no loss of resist thickness produces some undesirable exposure of 'unexposed' regions due to interference or scattering effects. If this dose exceeds D_n^i, which is often the case, a residue is left effectively broadening the image. It is usually avoided by exposing at a lower dose, e.g. $D_n^{0.7}$, where accurate reproduction of linewidths is obtained. This is the lithographically useful sensitivity which is usually quoted for the resist material.

It may be seen from Fig. 7.2 that the gel dose only approximates to a threshold and the degree to which this approaches a sharp threshold is known as the contrast. In a negative resist this is related to the rate of crosslinking and once more is dependent on the resist chemistry and polymeric parameters. It is most conveniently measured from the same plot of resist thickness with dose. It is usually defined as the gradient of the linear portion of the curve (Fig. 7.3)

$$\gamma = 1/(\log D_n^{1.0} - \log D_n^i)$$

where $D_n^{1.0}$ is the dose required for 100% resist remaining, extrapolated from the linear portion.

Contrast is independent of molecular weight but is strongly dependent on the distribution of molecular weights in a polymer.[4,5] Normally the polymer chains have quite a wide distribution of molecular weights and since the gel dose is dependent on weight for any given polymer the wider the distribution of weights within the polymer the wider the

S. Clements

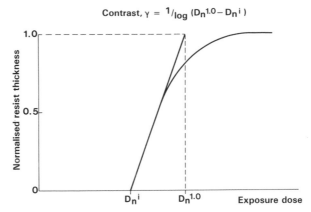

Contrast, $\gamma = {}^1/_{\log} (D_n{}^{1.0} - D_n{}^i)$

Fig. 7.3. Determination of contrast for a negative resist.

distribution of gel doses and the lower the contrast of the polymer as a whole.

Resolution is directly related to contrast, so high contrast is obviously desirable in a resist. Thus in high resolution lithography it may be desirable to reduce the polymer distribution by specialised preparative or fractionation techniques, although usually adequate contrast is achieved by correct choice of resist chemistry. In lower resolution lithography such as photolithography high contrast resists are still advantageous in obtaining steep profiles from cheaper lithography equipment with poor imaging. A reasonable contrast also helps overcome the problem of variation of dose with resist thickness. In photoresists the bottom of the film receives a lower dose than the top because of absorption by the upper parts of the film, whilst in electron resists the base of the film receives a heavier dose because of scattering. Consequently a slight overdose is required to ensure exposure of the entire resist layer.

It has already been observed that choice of solvent profoundly affects the sensitivity achieved in any given resist but few guidelines exist for this choice. Conventionally a marginal solvent is used, often a mixture of a solvent and a non-solvent, so a relatively small change in molecular weight renders the resist insoluble. It should be remembered that development is a kinetic process so temperature and other factors such as humidity are important if reproducible sensitivities are to be obtained, although the latitude in these factors is variable between resists.

7.2.1.2. Positive Resists

The sensitivity of positive resists is similarly determined by measurement of resist thickness with dose (Fig. 7.4). In this case the lithographically useful sensitivity is the dose at which the resist is just removed after development, D_p^0, without affecting the thickness of unexposed resist. The contrast is also determined as before as the gradient of the linear portion of the curve

$$\gamma = 1/\log{(D_p^0 - D_p^i)}$$

where D_p^i is the extrapolated dose at which 100% of the resist is left. In a positive resist the contrast is related to the rate of change of solubility with molecular weight which means that it is again dependent on the molecular weight distribution[6] although independent of molecular weight. It also means that it is strongly dependent on the developer solvent, so choice of developer is even more crucial than in negative resists. There also tends to be a greater change in sensitivity between solvents but there is virtually no change in sensitivity with molecular weight, in marked contrast to negative acting polymers.[7]

Since development in positive resist is achieved by differences in the rate of solution it is important to note that it is possible to 'force develop' the resist and achieve higher sensitivity by allowing the developer to remove some of the original unexposed resist.[8] Although higher sensitivity is produced by this method, the 'thinning' of the remaining resist results in problems of linewidth control and could mean higher defect densities especially in the form of pin-holes. In

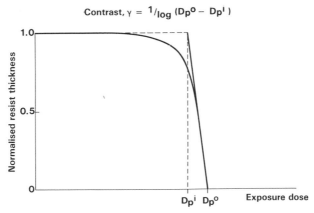

Fig. 7.4. Determination of contrast for a positive resist.

view of these effects it will be clear that care must be exercised in measurement or comparison of sensitivity and contrast data and their interpretation.

7.2.2. Resolution

The resolution capability of a resist is determined by the hardware used to expose it, its chemistry and the process in which it is used. Many of these parameters contribute to the resist contrast and the hardware determines, for example, the level of scattering or diffraction effects which impinge on the resist. These parameters take no account, however, of resist deformation which can occur during processing. Foremost amongst these deforming processes are thermal flow during baking and etch stages, and swelling on development. The susceptibility of a resist to these conditions usually determines the developing and baking steps used.

Swelling is a consequence of the solubility behaviour of polymers.[9] Dissolution occurs in two stages, first the solvent diffuses into the polymer matrix accompanied by a large increase in volume, and then, if the solvent–polymer interactive forces are sufficient to overcome the polymer–polymer interactions, the matrix is broken up to form a true solution. Swelling is the distortion of the resist image caused by the ingress of solvent into the insoluble regions of the resist during the removal of the rest of the resist in the developer. This may be manifested in the form of 'bridging' (Fig. 7.5(a)) where lines close together swell and join then shrink on removal of solvent, or a longitudinal buckling or 'snaking' (Fig. 7.5(b)) which arises from the constraint to swelling by adhesion of the resist film to the substrate.

Swelling is reduced in lower molecular weight polymers where there is less gel formation in advance of dissolution. This is a major reason why positive photoresists such as Shipley AZ containing low molecular weight novolac resin show such good resolution. In positive electron resists too, low molecular weight scission fragments can be dissolved by choice of appropriate developer solvent without significant swelling of unexposed regions. The resolution of negative resists, on the other hand, is usually limited by swelling, although higher resolution can be obtained at the expense of sensitivity through use of a lower molecular weight polymer.

Thermal flow occurs in a resist when processing is carried out around or above the glass transition temperature (T_g). Resist materials normally exist at room temperature in a glassy state in which molecular

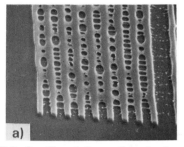
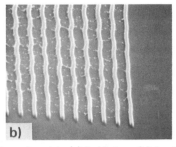

Fig. 7.5. Swelling deformations in developed resist: (a) bridging; (b) buckling. (Courtesy of Plessey Research (Caswell) Ltd.)

motion is severely restricted, but above the glass transition temperature some motion is possible and the polymer becomes rubbery and flows (Fig. 7.6). Ideally therefore processing should be carried out below the T_g, but in some processes such as plasma etching control is not possible so post-development treatments have been developed to process harden the resist. The most effective way of reducing polymer motion and raising T_g is to induce crosslinking. To this end flood deep UV[10] and Argon plasma treatments[11] have been developed for photoresists and electron resists and chemical treatment of novolacs with formaldehyde also produces crosslinks.[12] Hard thermal baking has also been used to the same effect.[13]

Resolution is usually assessed by examination of patterns under an electron microscope. Confusion can occur here because the measured resolution is different for different types of feature as a result of radiation scattering. Consequently isolated lines are the most difficult features to define in positive resists and isolated windows in negative resists. Assessment of resolution should therefore take account not only of processing details but of the feature being defined.

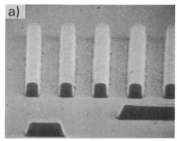

Fig. 7.6. Baking effects in an electron resist: (a) before; (b) after. (Courtesy of Plessey Research (Caswell) Ltd.)

7.2.3. Etch Resistance

The term etch resistance is somewhat ill-defined but is usually used in reference to the ability of the resist to withstand the process conditions during transfer of the resist pattern to the substrate. This includes not only its bulk chemical reactivity to etching reagents but also its adhesion to the substrate.

Adhesion is most critical with wet etching which is generally used for all but the finest lithography. Its assessment is difficult since it is dependent on factors such as wetting ability at the interface as well as straightforward adhesive force. Measurement of adhesion or use of 'sticky tape' tests therefore do not necessarily produce a relevant result. Normally, lack of adhesion is observed qualitatively by loss of linewidth control, peeling of features, or their partial or complete removal. Adherence of the resist is ensured in a number of ways. Primarily adhesion is aided by the presence of certain groups in the polymer structure, and hydroxy (–OH) or acid (–COOH) functionality has been used with success. Alternatively, although less desirable, adhesion is provided by promoting agents such as hexamethyl-disilazane, which bonds between the substrate and polymer. Whatever the resist, it is important to ensure cleanliness of the substrate surface and, in conjunction with a judicious post-development bake, this is often sufficient to obtain the required adhesion.

Dry etching techniques make fewer demands on adhesion but require higher thermal and radiation stabilities. This is particularly difficult in degrading resist systems such as positive electron resists, where materials are specifically chosen to be radiation-sensitive during exposure. In other resists too there is an increasing demand on dry etch resistance. An understanding of the chemical and physical properties of the plasma system is only just being gained, and so far only empirical guidelines exist for the design of resistant materials. From such studies[14,15] it appears that the presence of aromatic functionality or certain metals improves resistance to a range of commonly used plasmas. The effect of aromatic moieties is borne out by the difference in etch resistance between the rubber-based negative photoresists and the aromatic novolac positive resists. The use of metals is more selective. Organosilicon polymers[16] show great resistance to oxygen plasma as a result of reaction with this plasma to form the oxide, SiO_2. The formation of an oxide 'crust' effectively stops etching entirely. The use of other plasmas, however, will not produce this protective layer, so etch rates in, for example, halogenated plasmas are unchanged.

7.3. RESISTS FOR PHOTOLITHOGRAPHY

The majority of semiconductor processing is still carried out using photolithography with exposure over the 350–450 nm wavelength range of the near ultra-violet. The materials used were inherited from other industries[1] yet were found to be readily adaptable to the new demands of semiconductor device manufacture. Only recently, as optical lithography has been pushed closer to the fundamental limitations of wavelength, has it proved necessary for greater refinement and better understanding of the chemistry of the photoresists. The most obvious restriction of photoresists is the limited energy available in the ultraviolet to initiate the exposure reactions. This limitation led to the use of two-component photoresists where in simple terms a photosensitive component interacts with a polymeric matrix. In this way exposure characteristics are separated from etch resistance and film-forming properties. This approach has the added advantage that exposure is selective, in contrast to high energy lithography where indiscriminate bond-forming and bond-breaking reactions occur.

7.3.1. Negative Photoresists

Negative optical resists were, until recently, the most used resists in microlithography, responsible for the manufacture of all kinds of integrated circuits and devices. Although the components vary in detail between manufacturer, the underlying chemistry in, for example, Hunt NMR, Kodak KTFR and Merck Selectilux-N remains the same. In each a partially cyclised rubber is crosslinked on exposure by bis-azido compounds.[17] Originally the rubber used was natural but impurities led to batch-to-batch control problems and now synthetic rubber is used, prepared by Ziegler–Natta polymerisation of isoprene. Poly(*cis*-isoprene) is, without further treatment, too soft a binder material for use in lithography, having all the problems connected with a low glass transition temperature. Partial cyclisation (Fig. 7.7), however, produces a hard polymer with good adhesion and film-forming properties. This cyclisation process must be well controlled to produce the required degree of modification and reproducibility, and largely determines the success of the resist.

The azide photosensitisers used in negative resists show good thermal stability and readily decompose with high efficiency on irradiation. Whilst individual azides have relatively narrow absorption bands (Fig. 7.8), molecular tailoring allows a range of compounds to be

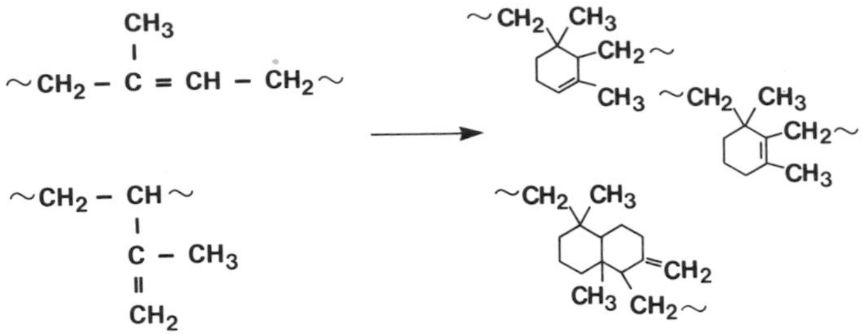

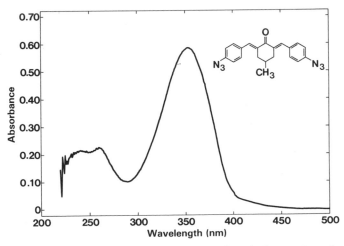

Fig. 7.7. Cyclisation of polyisoprene.

produced with absorptions covering wavelengths of 300–450 nm. Thus sensitisers can be chosen with selective absorption of any particular mercury lines used by exposure systems. Exposure of the azide produces a highly reactive nitrene species which undergoes insertion reactions with carbon–hydrogen bonds and carbon–carbon double bonds (Fig. 7.9). These are plentiful in the cyclised resin so it is clear that the two azide groups on the sensitiser can readily produce a crosslink. A network of such crosslinks reduces the solubility in the organic developer enabling an image to be produced.

Fig. 7.8. Structure and ultra-violet spectrum of typical negative photoresist sensitiser.

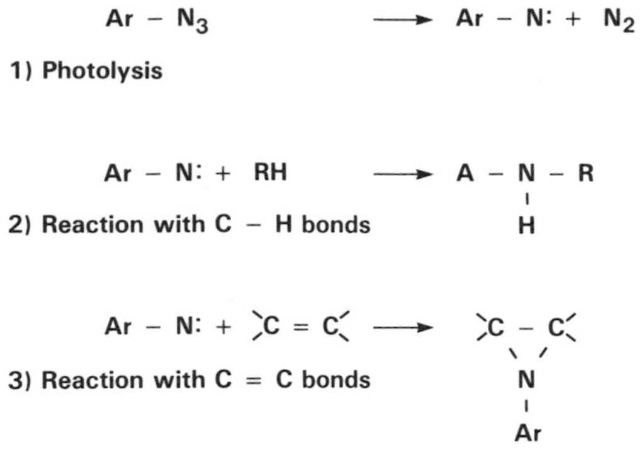

Ar – N$_3$ ⟶   Ar – N: + N$_2$

1) Photolysis

Ar – N: + RH   ⟶   A – N – R

2) Reaction with C – H bonds |

                                  H

Ar – N: + >C = C<   ⟶   >C – C<

3) Reaction with C = C bonds   \ /

                                  N

                                  |

                                 Ar

Fig. 7.9. Insertion reactions of photosensitisers.

7.3.2. Positive Resists

Whilst negative photoresists were the mainstay of the semiconductor industry they are now being increasingly replaced by positive resists which operate in a more involved manner and to some extent are still not entirely understood. Their advantages, however, are quite significant now that the technology is being extended to its limits. Positive resists show higher resolution, superior etch resistance in wet and dry environments, and better thermal stability, especially when given appropriate 'hardening' treatments. In negative resists crosslinking is readily induced using sensitisers operating in the near ultra-violet but the energy available in photons at these wavelengths is insufficient to induce bond scissions in any normal polymer, hence a more complicated mechanism is used to enhance the solubility.

In brief, the 'sensitisers' used in positive photoresists are large photoactive compounds which are insoluble in basic solvents and sufficiently bulky to inhibit solubilisation of a base soluble polymeric resin. The action of light breaks down the photoactive compound to a base soluble product and the whole exposed region may be developed out. The particular strength of the photoresists, however, arises from the subtleties of the chemistry both on exposure and during development.

As in negative photoresists there are many suppliers who produce positive acting materials which employ more or less the same chemistry but in slightly different formulations. The polymeric resin compound is usually a novolac resin (Fig. 7.10(b)) of fairly low molecular weight,

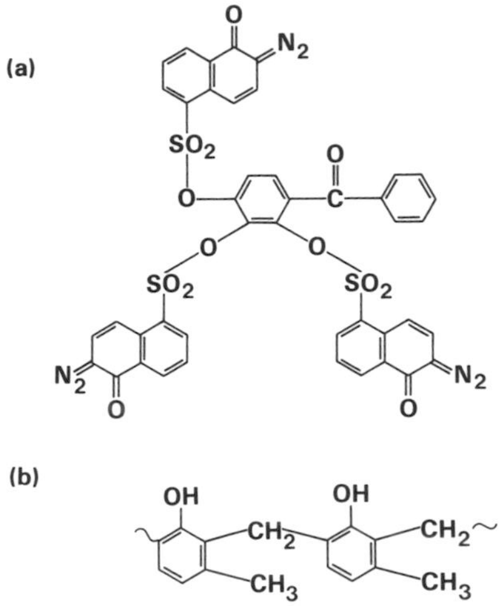

S. Clements

(a)

(b)

Fig. 7.10. Typical components of positive photoresist: (a) substituted diazonaphthoquinone sensitiser; (b) phenolic resin.

typically between 1000 and 2000, making them readily soluble in alkaline solutions. Once again it is the polymeric resin which is responsible for the film-forming characteristics and tendency to flow or otherwise during processing. The dissolution of the resin is inhibited by a large molecule which is invariably a naphthoquinone diazide derivative (Fig. 7.10(a)). Decomposition of the diazide occurs in two stages under normal conditions ultimately yielding a carboxylic acid (Fig. 7.11). The acid product means that the inhibitor is now soluble in alkali developer and dissolution of the exposed resist is possible.

During development there is also found to be an apparent decrease in the solubility of the unexposed regions, over and above that produced by the inhibitor alone.[18] It is thought that crosslinks are produced between the resin and unreacted photoactive compound in the presence of the alkaline developer (Fig. 7.12). Hence not only is there

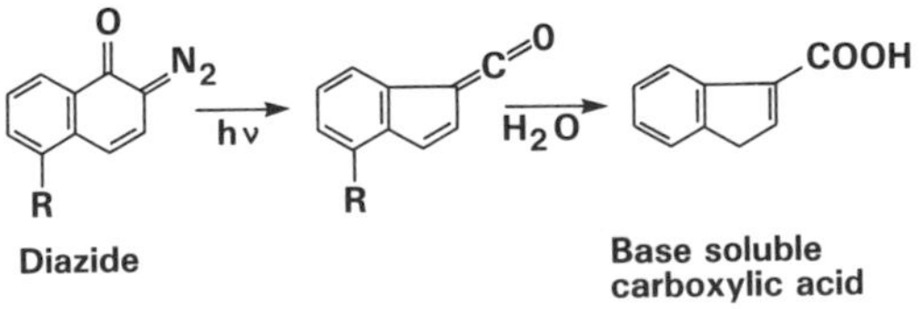

Diazide   Base soluble carboxylic acid

Fig. 7.11. Photolysis of diazide photosensitiser.[18]

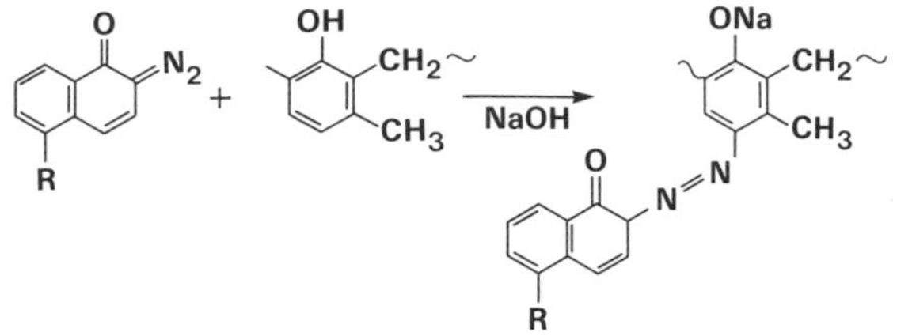

Fig. 7.12. Proposed reaction of unexposed regions in developer.

an enhancement of the solubility of the exposed resist but there is also a concomitant reduction in the solubility of the unexposed regions. This goes some way to explaining the excellent contrast seen in positive photoresists.

It is important to note that water is essential in the exposure of the resist. Although sufficient water is normally present as an impurity in the hydrophilic resin and is readily picked up from the air and the spinning solvent, control of humidity is still necessary for reproducible working. Deliberate exclusion of water produces an entirely different set of reactions and crosslinking occurs with the novolac resin. This has been used to reverse the tone of the image by employing a flood exposure in vacuum after masked exposure in air.[19] In general, however, it is found that this reaction is not so well controlled. A better controlled process, the 'monazoline process', has been developed which produces a negative image by a different mechanism.[20] After normal exposure the resultant acid product is decomposed by baking in the presence of an amine additive (normally a monazoline), with the consequence that the exposed resist remains insoluble. Flood exposure of the whole resist then renders the previously unexposed resist soluble and a negative image is produced on development. Without the bake and flood exposure the monazoline-doped resist develops a positive image in the normal way.

7.3.3. Exposure Techniques

So far, consideration has been given largely to the chemical factors determining the resist performance. The exposure method and quality of the hardware, however, are fundamental in dictating the type of resist, its many requirements and the quality of images produced. One aspect of this is the chosen irradiation wavelength.

In photolithography, using 300–450 nm ultra-violet, high-pressure

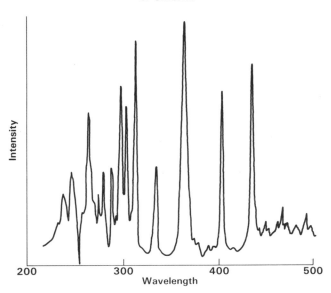

Fig. 7.13. Spectral output of high-pressure mercury/xenon lamp.

mercury or mercury/xenon lamps are routinely employed. The mercury emission in this range consists of several particularly strong lines (Fig. 7.13) which are responsible for exposing the resist. According to the method used some or all but one line may not be transmitted because of deliberate filtering, absorption or defocussing by the optic system. The sensitisers must therefore be tuned to make use of what emissions are transmitted.

7.3.3.1. Contact and Proximity Printing

The original techniques employed for exposure of resists used simple shadow printing techniques where the resist-coated substrate was flood-exposed with ultra-violet light through a patterned mask which was close to, or in intimate contact with, the resist surface. These systems are still used for production of small numbers of devices where their simplicity and cheapness are a great advantage. They also make best use of the whole spectral output of the irradiation source which is normally limited only by the absorption of the mask substrate material.

Contact printing also gives the best reproduction of the mask pattern, in favourable conditions, and produces the highest resolution possible in photolithographic techniques. The theoretical resolution limit for a mask grating with equal sized lines and spaces is given by

$$b_{min} = 3/2(\tfrac{1}{2}t\lambda)^{1/2}$$

where b is the wdith of the lines, t the resist thickness and λ the irradiating wavelength. This theoretical limit, however, is not usually achieved in practice for a number of reasons and has been superseded on the whole by other methods. One problem is in obtaining the intimate contact with the resist; mask and substrate flatness across the entire surface is not possible, and debris between the two causes further deterioration in resolution. It is this debris which is the major problem in contact printing. The particles give rise to defects in the patterned substrate and more importantly damage the patterning mask which is reproduced in subsequent exposures. Not only does this limit the lifetime of expensive masks but quickly diminishes yields of patterned substrates.

These shortcomings led to the use of proximity printing where the mask was deliberately kept out of contact with the resist, accepting the loss in resolution caused by the increase in diffraction effects (Fig. 7.14). The theoretical limit for resolution is a simple extension of the equation for contact printing

$$b_{min} = 3/2[\lambda(s + \tfrac{1}{2}t)]^{1/2}$$

where s is the gap between mask and resist. This improves mask lifetime and device yields but resolution is limited by how small a gap can be safely used. This means using wafers and masks of well-controlled flatness and, in practice, a 10 μm gap is probably the minimum that can be used.

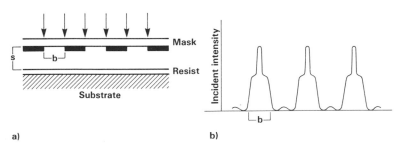

a)   b)

Fig. 7.14. (a) Schematic of proximity printing. (b) Resultant intensity incident on the resist because of diffraction effects. (Reprinted by permission of The Electrochemical Society, Inc., Ref. 21.)

7.3.3.2. Projection Printing

Advances in lens optics and economics of scale permitted the development of projection printing as a means of overcoming the shortcomings of shadow printing. This technique further sacrifices resolution as a result of lens aberrations and limitations but in return higher yields are achieved through lower defect densities and prolonged mask life which , permits high-quality masks to be used. In addition, alignment accuracies can be improved, an increasingly important factor in complex circuits currently being developed.

Two generalised systems have been developed using either reflective or refractive optics. The reflective systems, exemplified by the Perkin–Elmer Micralign, provide the main exposure systems in many large-scale semiconductor production processes. Their high throughput capability, of maybe 50 wafers an hour, made possible the abundance and relative low cost of home micro-computers. Refractive systems for large area imaging require complex and expensive lens optics but the better resolution and registration accuracy over reflective optics has led to their use for fine line photolithography at less than 2 μm.

The resolving power of projection optic systems is described by the concept of the modulation transfer function (MTF).[21] Diffraction of light at the edge of a mask feature leads to a blurring of the image which for a fine grating results in a considerable intensity of light in otherwise unexposed regions (Fig. 7.15). The modulation is defined by

$$M = \frac{I_{max} - I_{min}}{I_{max} + I_{min}}$$

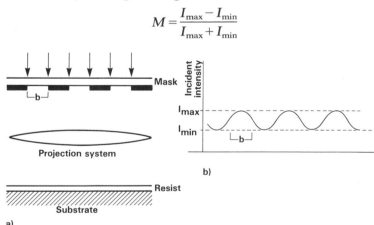

Fig. 7.15. (a) Schematic of projection printing system. (b) Resultant intensity incident on the resist because of modulation by the projection system. (Reprinted by permission of The Electrochemical Society, Inc., Ref. 21.)

where I_{max} and I_{min} are the intensity maximum and minimum in the exposed and unexposed regions of a grating, respectively. The MTF is then defined as the ratio of the modulation in the mask plane (idealised transfer profile) to that in the imaged plane (actual transfer profile).

The resolution capability as measured by the MTF is largely determined by the numerical aperture of the optical system and the exposure wavelength

$$\text{Minimum feature size} \propto \lambda/\text{NA}$$

The first-order dependence with wavelength, compared with just a half-order dependence in shadow printing techniques, leads to a big incentive in using shorter wavelengths. In reflective systems the use of shorter wavelength radiation is limited only by lens aberration since, unlike refractive systems, reflective optics focus all wavelengths equally. Improved resolution has been demonstrated with reflective optics which use the 310 nm mercury emission.[22] Refractive systems suffer from fabrication problems for quality optics at shorter wavelengths and are currently limited to radiation of over 400 nm. Higher numerical aperture, however, means that refractive optical systems have overall better resolution than reflective projection systems. Highest numerical aperture is achieved with reduction lenses, and 10:1 and 5:1 reduction optics are now commonly used in wafer steppers. The improvement in numerical aperture is obtained only at the expense of the image field size and it is because of the reduced field diameter that the step-and-repeat approach is used. This reduces wafer throughput but for high-resolution applications wafer steppers are currently the foremost technology.

Apart from optic problems a limit to higher numerical aperture systems is set by depth of focus considerations. Whilst larger numerical apertures improve resolution, a concurrent reduction in depth of focus is seen. The depth of focus, δ, as derived from the Rayleigh criterion, shows an inverse square dependence on the numerical aperture

$$\delta = \frac{\lambda}{2(\text{NA})^2}$$

which becomes increasingly important with an increased circuit complexity and consequent non-planar topology. It is for this reason that further advances in resolution in optical lithography will turn to shorter wavelength radiation rather than larger numerical aperture.

S. Clements

7.4. ELECTRON LITHOGRAPHY

There is no doubt that photolithography is used to produce the vast majority of devices and will continue to hold that position for many years. The trend to ever smaller geometries, however, has led to increasing use of other techniques employing higher energy radiation, the foremost of which is electron lithography. The primary advantage is the shorter wavelength giving an inherently finer resolution which is being increasingly used for definition of masks and reticles for photolithography.

Conventionally, electron-beam exposure is carried out using a 'pencil' of focussed electrons to write the pattern using a beam deflector driven by a computer system. The beam in this case is scanned across the substrate to be exposed using either vector scan or raster scan methods (Fig. 7.16). Vector scanning writes patterns in a similar way to which shapes might be filled in with a ball-point pen whilst the raster uses a system similar to that of a television tube where the entire surface is scanned and the beam turned on and off as necessary. Each method has advantages; vector scan machines have to scan smaller areas but this is compromised by the need for a more sophisticated deflection system. The raster scan by contrast, whilst having to irradiate the entire substrate, can be operated at faster scan speeds. Normally the vector scan uses less exposure time when less than around one sixth of the area has to be exposed.

Exposure times by either method are inevitably much slower than the flood exposure through masks used in photolithography, so throughput and running costs are higher. Some research into analogous image projection systems is being carried out to make it more economic and beam shaping methods are also being examined as a way of improving exposure time.[23]

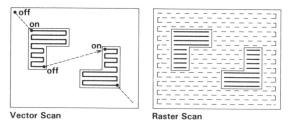

Fig. 7.16. Scanning methods in electron beam lithography.

One advantage of using a beam system is the ability to expose substrates directly without a mask. Although direct writing is considerably slower it has several increasingly important applications. When devices or circuits are only being produced in small numbers, direct writing avoids the need to produce a mask set and computer-aided design data can be used by the system data to drive the electron beam.

Another use of direct writing is in hybrid lithography[24] where the majority of the circuit is patterned by photolithography and the high-resolution capability of the electron beam is used to produce just the small geometries. An extension of this approach has been adopted for electron-beam customisation of mass-produced uncommitted logic arrays.[25] These applications in addition to mask-making will ensure the continued importance of electron lithography in the future.

The major difference between photolithography and electron lithography when considering the types of resist material involved is the energy of the impinging radiation. The energy associated with an electron accelerated by 20 kV is several hundred times that required to break chemical bonds and hence any bond has a possibility of being broken. Most bonds broken in such a way will undergo relaxation or recombination but scission and crosslinking are generated during exposure with one prevailing over the other depending on the conditions and particular chemistry of the material involved. The general approach in high-energy lithography is to use polymers inherently susceptible to scission or crosslinking. In the case of negative, crosslinking electron resists there is a host of groups which might be employed but in practice there is a limit set on the most sensitive by their thermal stability. The reactions initiated in electron-sensitive polymers also tend to be initiated by thermal reactions and high sensitivity results in significant reaction at room temperatures with consequent short shelf-life. The same limit exists in positive resists but this usually becomes apparent after development where a high thermal stability may be required in the unexposed image during processing. Ways of overcoming this problem will be discussed later.

The wavelength of the electron is of the order of a hundredth of a nanometre so resolution is no longer diffraction-limited. The electrons are, however, easily scattered by atoms and molecules, and scattering within the polymer produces a broadening of the written line. In the high-energy beams of around 20 kV typically used in lithography the scattering not only occurs within the resist but also back from the substrate and contributes significantly to the dose experienced by the resist. This contribution also varies of course between substrates increasing with the molecular weight of the substrate. It has been shown by Monte Carlo simulations that at an accelerating voltage of 20 kV on a silicon substrate the back-scattered electrons are spread over a distance of 4 μm.[26] These many factors influence the particular requirements of electron resist materials; high sensitivity must be combined with thermal stability and high contrast is required to discriminate the exposure dose from back-scattering.

7.4.1. Positive Electron Resists

Since the actual chemistry involved in determining the electron sensitivity of a resist is complex, a more general approach has been adopted. The resist sensitivity is related to a polymer-dependent constant G_s defined as the number of chain scissions per 100 eV of absorbed energy. This is derived from the molecular weights of exposed and unexposed resists

$$\frac{1}{M_n^*} = \frac{1}{M_n^0} + G_s D / 100 N_a$$

where M_n^* and M_n^0 are the molecular weights of exposed and unexposed resist, N_a is Avogadro's number, and D is the exposure dose.[27] In order to obtain sufficient polymer to analyse, the polymer is most conveniently irradiated by gamma radiation but results are found to be consistent with those derived by longer methods using electron beam exposure.[28] The correlation between G_s and exposure sensitivity is usually very good although some variation is apparent because of the effects that different processing has on resist sensitivity. G_s accounts only for the exposure sensitivity excluding any ability of the solvent to differentiate.

The first polymer studied as an electron resist was polymethyl methacrylate which is a classical degrading polymer. It has high resolution and, despite low sensitivity and poor dry etch resistance, is still used in some processes and is widely used as a standard against which other materials are compared. The low sensitivity of 100 μC cm^{-2} is borne out by a G_s value of 1·3. Because of its simplicity and wide availability it is probably the most studied resist material and its degradation mechanism has been well characterised.[29,30] A full discussion of the mechanism is not necessary here, but in general terms the degradation is initiated by cleavage of the side groups leading to fragmentation of the polymer chain. Attempts to improve the relatively poor sensitivity of PMMA have therefore centred on effecting this scission process more readily by substitution of different side groups. A large research effort has produced many acrylate analogues some of which have been adopted in production processes. One such resist is 'terpolymer'[31] a polymer produced from a copolymer of methyl methacrylate and methacrylic acid (Table 7.1). On baking the copolymer a proportion of methacrylic anhydride is produced resulting in a resist with a G_s value of 4·5 and a sensitivity of 10 μC cm^{-2}.

Substitution of the side chain with fluorine was shown to be beneficial in improving sensitivity and FBM,[32] polyhexafluorobutyl methacrylate and EBR-9,[33] polytrifluoroethyl α-chloromethacrylate are produced commercially with sensitivities of 1–5 μC cm^{-2}. Unfortunately most PMMA analogues sacrifice some other property to achieve such sensitivities, e.g. FBM has poor adhesion and its process latitude is small.

Another shortcoming of PMMA in semiconductor processing is its low thermal stability to flow. It was stated earlier that a most effective

Table 7.1
Positive electron resists

Resist Material	Repeat Unit Structure	G_s	Sensitivity
PMMA	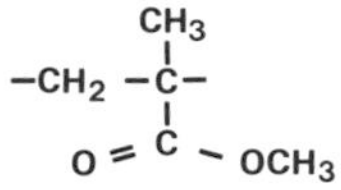	1.3	100μC cm^{-2}
Terpolymer	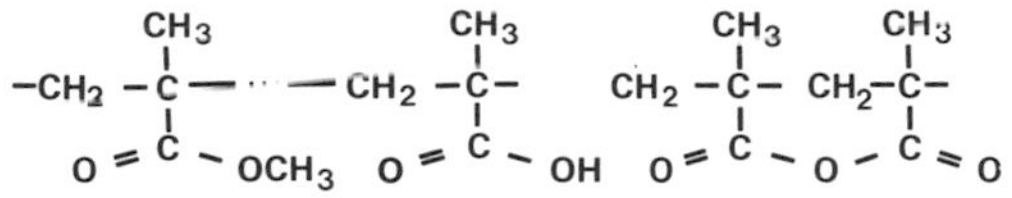	4.5	10μC cm^{-2}
FBM	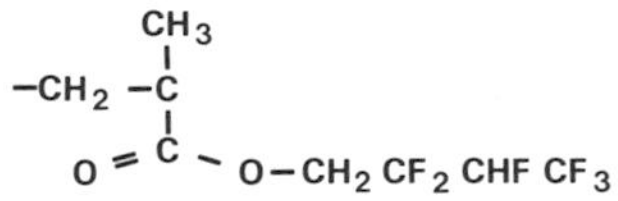	5.0	0.4μC cm^{-2}
EBR-9	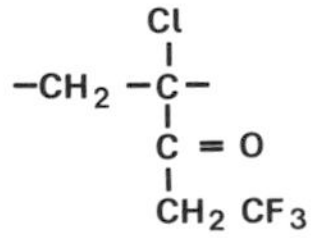	–	3μC cm^{-2}
PBS	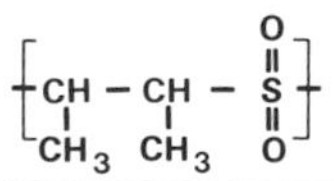	10.0	0.7μC cm^{-2}

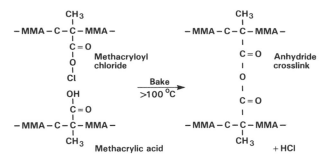

Fig. 7.17. Crosslinking on baking methyl methacrylate copolymer resist (see text).

way of increasing T_g and hence flow stability is to induce crosslinking. This approach has been used in PMMA to produce thermal stability of over 200°C.[34] The crosslinking is induced by baking the resist film which contains small amounts of methacrylic acid and methacryolyl chloride (Fig. 7.17). These react on baking to form anhydride cross-links which are degraded like the rest of the polymer on exposure. An added advantage is that by varying the bake temperature the degree of crosslinking can be varied, with a consequent effect on the exposure dose which can be as low as 20 $\mu C\, cm^{-2}$.

An entirely different approach to high sensitivity in positive resists is represented by the polyolefin sulphones.[35] Their high scission value, G_s, of 10 derives from ready depolymerisation or 'unzipping' on exposure back to the constituent monomers, sulphur dioxide and the olefin. Polybutene sulphone with a sensitivity of 1 $\mu C\, cm^{-2}$ has for a long time dominated mask-making applications where speed is most important in ensuring high throughput. However, for direct-write applications and dry etch masks its poor dry etch resistance and thermal stability prohibit use.

Conventional positive photoresists were, at one time, commonly used as electron resists, being readily available in high quality and with relatively well-characterised properties. On exposure to electrons, positive images are produced but a different series of reactions occur[36] and crosslinking can be induced, similar to that produced by ultra-violet exposure in vacuum. This reduces the resist contrast and at doses only slightly higher than that necessary for normal exposure actually produce negative images. These disadvantages arise from the unsuitability of the diazide sensitiser. Replacing the ultra-violet sensitiser with an electron-sensitive polyolefin sulphone, however, has been successfully used to produce a positive resist which combines the etch resistance and processability of the novolac resin with the sensitivity of the polyolefin sulphone.[37]

7.4.2. Negative Electron Resists

Materials developed for use as negative electron resists fall into fairly well-defined families based on a limited number of reactive crosslinking groups. Epoxide and allyl groups have been employed with success in resists of high sensitivity. The epoxide group combines high sensitivity with thermal stability and good adhesion to common substrates, but displays poor resistance to dry etching media and a persistence of reaction after exposure which can affect linewidth control. A copolymer of glycidyl methacrylate and ethyl acrylate, COP,[38] is widely used in mask-making because of its sensitivity of $0\cdot8\ \mu C\ cm^{-2}$ (Table 7.2).

Resists containing reactive vinyl and allyl groups are generally found to suffer shelf-life problems but some resists, notably those containing allyl methacrylate,[39] have been produced commercially with adequate processing properties.

The desirability of dry etch resistance especially in direct-write applications has led to the development of new resists, largely based on polystyrene. Polystyrene has an etch resistance in most plasmas comparable to positive photoresists but its poor sensitivity of around $100\ \mu C\ cm^{-2}$ precludes its use in most applications. However, it was found that substitution of the benzene ring with chlorine or certain other halogenated groups enhanced the sensitivity by more than an order of magnitude, without a significant effect on etch resistance. Copolymers of chlorostyrene have been produced with styrene[40] or glycidylmethacrylate.[41] The enhanced sensitivity is thought to derive from the more labile carbon–halogen bond in comparison with the carbon–hydrogen bond, to enable crosslinking via the aromatic side chain. A further improvement was expected and indeed found when chloromethyl groups were used in place of simple chlorine. Polychloromethyl styrene and various combinations of this resist have been produced by several workers[42–44] and shown to combine high sensitivity and etch resistance. The resolution of all these resists is limited by swelling during development and work is currently being devoted to producing resists which reduce or overcome this problem. One approach is to lower the molecular weight of the resist.[45] This reduces

Table 7.2
Negative electron resists

COP	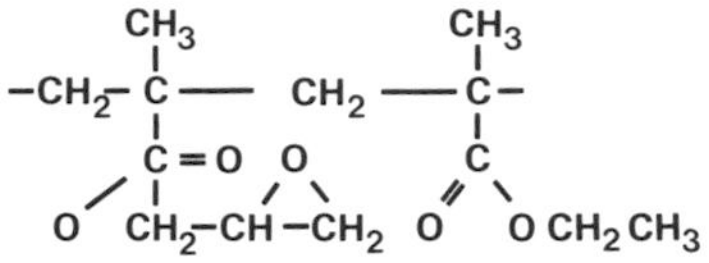 CH3 CH3 –CH2–C— CH2 —C– C=O O C O CH2–CH–CH2 O O CH2CH3
Polystyrene	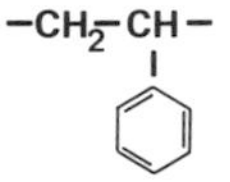 –CH2–CH–
Polychlorostyrene -co-styrene	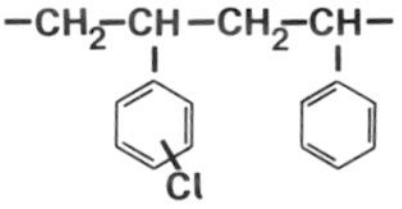 –CH2–CH–CH2–CH– Cl
GMC	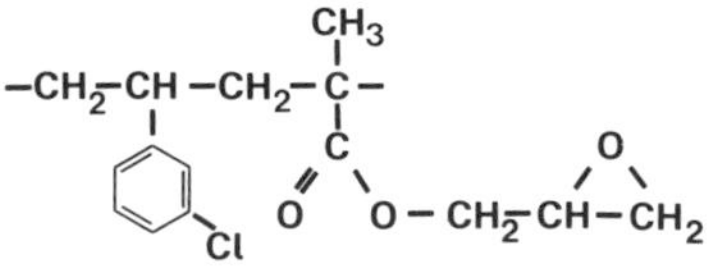 CH3 –CH2–CH–CH2–C– C O O O O–CH2–CH–CH2 Cl
Polychloromethylstyrene	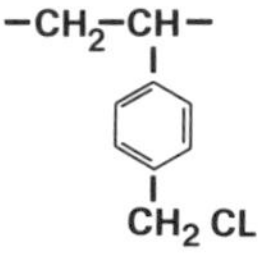 –CH2–CH– CH2 CL
Organosiloxanes	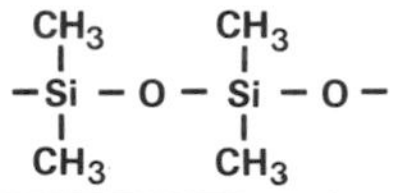 CH3 CH3 –Si – O – Si – O – CH3 CH3

swelling but also lowers the sensitivity which means materials with ever higher sensitivity are required. An alternative method is to find materials which show an inherent low tendency for swelling. Organosiloxanes have been reported to show high sensitivity yet little swelling and consequently have high resolution.[46] A third option which is now being examined is to change the chemical nature of the resist on exposure rather than inducing crosslinking, so that developer solvent only penetrates the region being removed.

One resist which uses this idea induces the change by a charge-transfer mechanism.[47] A strong electron donor, tetrathiafulvalene (TTF), bonded to a polystyrene backbone (Fig. 7.18) donates an electron to carbon tetrabromide on exposure. The exposed regions are therefore ionic in nature and readily dissolved in polar solvents whilst

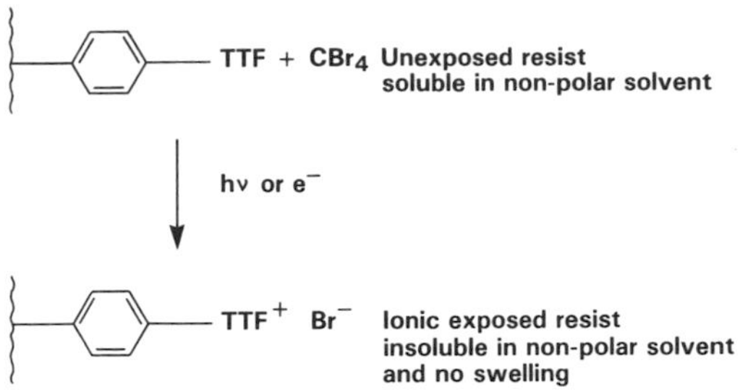

TTF + CBr$_4$ Unexposed resist
soluble in non-polar solvent

hν or e$^-$

TTF$^+$ Br$^-$ Ionic exposed resist
insoluble in non-polar solvent
and no swelling

Fig. 7.18. Non-swelling negative resist.

the unexposed resist can be developed in non-polar solvent. In this way positive or negative images may be produced by suitable use of developer but without mutual swelling of remaining material. Other new schemes exploiting polar/non-polar regions have been reported[48] and it is likely that further progress will be made in this area in the future.

7.5. X-RAY AND ION-BEAM LITHOGRAPHY

It has already been noted that in determination of radiation scission values for electron resists, gamma radiation could be readily substituted and a good correlation existed. It is found that this is generally the case for all electron resists, both negative and positive,[49] and is also found to extend to their sensitivity to ion-beams.[50] It is apparent from this that the resist chemistry must be virtually the same for all high-energy radiation and that the resist materials are interchangeable between technologies.

Whilst the correlation is very good some anomalies do arise. The sensitivity of a resist is partly determined by the absorption coefficient to the radiation. Just as photosensitisers are tuned to wavelengths so resists show varying absorptions to higher-energy radiation. This results in an overall ten-fold increase in the sensitivity of electron resists when exposed to ion-beams because of their greater absorption cross-section compared to electrons. In X-rays the absorption tends to increase with atomic weight, but actual values vary with the wavelength of the X-ray emission used. For example, the 4·3 and 4·6 Å wavelengths emitted by the Pd and Rh sources are strongly absorbed by bromine, and brominated resists have been successfully developed for enhanced sensitivity.[51]

X-ray resist materials, however, must not only be strongly absorbing but also have an inherent reactivity. Ferrocene, because of its iron content, absorbs over a wide X-ray range but is not very reactive and its sensitivity is low. The most promising resists have therefore been developed by modification of materials developed for electron lithography, incorporating absorbing groups. The principles involved are otherwise the same as those described for electron resists.[52]

7.6. DEEP UV LITHOGRAPHY

Whilst X-rays are being examined for possible large volume, fine resolution lithography to replace conventional photolithography an obvious alternative is the use of shorter wavelength ultra-violet. Already the advantage of using 310 nm radiation has been discussed but use of deep ultra-violet at less than 300 nm is now being developed. Deep UV lithography has the benefit of being simply a progression of current technology rather than an entirely new technology and maintains the specificity of resist chemistry by use of selective chromophores, enabling a better controlled process. Even so, several problems must be overcome relating to hardware, including lens optics and radiation sources, and resist materials.

The difficulties associated with lens optics concern the availability of optical materials with differing refractive indices which are also sufficiently transparent below 300 nm. At present, in the absence of such materials, reflective optics must be used. More fundamentally, if processing is to prove economic, brighter sources are required. Conventional mercury sources have a strong emission at 254 nm but this undergoes efficient self-absorption so overall brightness is low. One promising alternative being developed is the use of excimer lasers.[53] Normally lasers are not suitable for use as sources because of their coherence which gives rise to speckle from the constructive and destructive interferences of wavefronts. Excimer lasers, however, are not coherent and are capable of delivering several watts of power over a range of wavelengths. With this level of power high-sensitivity resists would not be necessary since exposure time would be negligible and emphasis could be placed on resolution, film-forming properties and etch resistance.

Materials presently used for deep UV lithography are exposed with mercury sources so sensitivity remains important. Near UV photoresists cannot be used below 300 nm because of absorption by the polymeric resins and, in general, electron resists have been employed. Polymethyl methacrylate (PMMA) was one of the first materials used as a deep UV resist, having, in common with many electron resists, an absorption in the 200–300 nm range. PMMA is positive acting but its sensitivity is poor and new materials incorporating indenone[54] and oximino-butanone[55] as copolymers have been reported. Some negative electron resists including chlorinated styrene derivatives have been reported for use as negative acting deep UV resists,[56] but swelling once more limits their resolution.

Other new sensitisers are being developed for both negative[57] and positive resists[58] but the potential of such resists will be strongly influenced by the progress in exposure hardware. Deep UV lithography will, however, be a promising contender for future sub-micron lithography.

7.7. DRY DEVELOPED RESISTS

The growing trend to use plasmas created a need for resists which were compatible with dry etching but it also increased interest in their possible use in producing the primary resist image. Most process steps including ion implantation, deposition and resist stripping can already be carried out without the need for wet processing, and dry development is seen as a natural extension of this.

The use of plasmas to develop the images has several potential advantages over solvent development:

(1) An improvement in resolution because of the elimination of solvent swelling effects.
(2) An expected reduction of defects and consequent improved yield as a result of handling in vacuum.
(3) An ability to automate the process to improve reproducibility.
(4) A reduced environmental problem associated with the disposal of solvent waste.

In order that the image may be developed in a plasma the exposure must effect a change which improves or reduces the etch rate of the material. However, conventional crosslinking or chain degradation

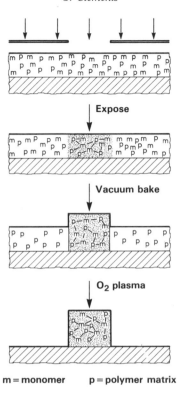

m = monomer   p = polymer matrix

Fig. 7.19. Plasma developed resist.

alone is not sufficient to effect this change and a more involved process is required. A generalised scheme which has been employed involves the incorporation of an etch-resistant compound within an otherwise plasma labile polymer. Where the etch-resistant compound is not bound into the polymer on exposure it is volatilised in a vacuum bake (Fig. 7.19). It will be remembered that etch-resistant materials include aromatic or organometallic compounds and both have been tried. Low molecular weight is also required in order that it may be easily volatilised and of course it must be radiation-sensitive.

A resist which is developed in this way has been reported for X-ray exposure[59] employing the plasma-labile poly dichloropropyl acrylate and an organosilicon monomer which binds through its acrylate functionality on exposure. Other schemes using aromatic monomers have been reported which are exposed by the ultra-violet.[60,61]

Effectively the same system has been used by other workers[62] employing the negative photoresist sensitisers the bisarylazides. Instead of volatilising during a bake stage, however, this apparently undergoes a reaction on baking which renders the exposed region plasma-resistant. Good differential etch rates have been reported.

To be more widely accepted for use in processing these resists still require improvement. Edge acuity and contrast are often lacking and process latitude is poor. This largely arises from the less discriminatory power of the plasma in comparison to dissolution in wet developers but work is continuing on new systems.

7.8. MULTILAYER RESISTS

The numerous requirements of the resist in any technology have been amply demonstrated and those of sensitivity, resolution and etch resistance have been described in detail. It is apparent from this that many of the demands directly conflict and one property is often sacrificed in order to optimise another. For example, resists capable of high resolution commonly require extended exposure times, and choice of molecular weight must compromise increased sensitivity with increased swelling on development. Another problem concerns the resist thickness. Thinner resists produce a proportional improvement in resolution but also reduce their ability to protect the substrate in etching media.

The successful photoresist materials benefit from dividing the roles of the resist between two components. The polymeric resin endows the necessary etch resistance, adhesion and film-forming properties whilst the sensitiser is largely responsible for the resist's sensitivity. Even so, the photoresists are being driven to their limits in terms of resolution over more and more complex topographies which makes it ever more difficult for any single resist film to meet all the requirements.

These limitations led to the important development of multilayer resist schemes where, rather than using more than one resist component, the resist properties are divided between two or more layers of resist. Since their first use in the early 1970s many schemes have been developed, but typically the multilayers consist of a thin top resist layer, which is exposed and developed conventionally, on top of a thick adhering bottom layer which is ultimately used to pattern the substrate. Although not always necessary, an intermediate layer (or layers) is often used to transfer the upper image into the base layer (Fig. 7.20).

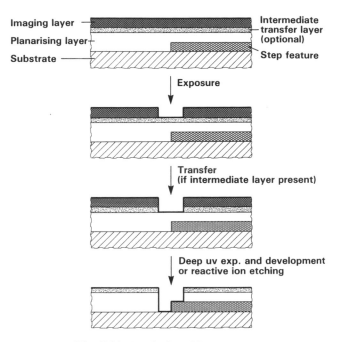

Fig. 7.20. Typical multilayer scheme.

The advantages of this system are many. Primarily, improved resolution and linewidth control are achieved from the reduced imaging layer thickness, made possible by the thick underlying layer. The base layer, which is usually PMMA, polyimide or positive photoresist, may be 2 or 3 μm thick. In a single resist layer thickness variations produced in going over a step lead to varying linewidths after development. Planarisation overcomes this problem and also reduces the effects of reflections from the substrate in photolithography or back-scattering in beam lithography. A further benefit is the reduced demand on depth of focus in a planar imaging resist layer, a particular benefit in short wavelength UV systems.

Two techniques are principally used to transfer the upper image into the base layer, flood ultra-violet exposure[63] and anisotropic reactive ion etching (RIE)[64] although wet processing is also used. Flood exposure is used in conjunction with positive resist base layers using the upper layer as a mask followed by development. The use of RIE requires some resistance of the thin top layer so intermediate layers are often used in this case. The intermediate layer is normally a thin inorganic layer typically imaged by selective etching through the upper mask before transferring by oxygen RIE to the base polymer layer. In this way the minimum of etch resistance is required in the upper layer. The role of the top layer in these applications has led to use of the term 'portable conformable mask' or PCM.

Another use of multilayer resists is in profile control. The PCM-type multilayer system is largely used for high-resolution applications with good linewidth control over complex geometries, but a further use of the multilayer system is in profile control. In this case isotropic etches are used which produce undercut or overcut profiles through suitable differential etch rates between layers. A common variant of this is two resists which are developed in mutually exclusive developers;[65] the control imparted by the multilayer system has found particular application for float-off metal deposition (Fig. 7.21).

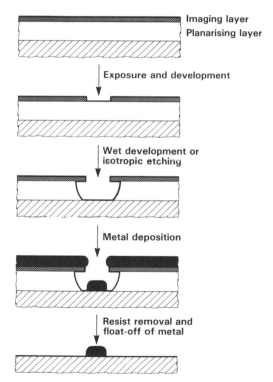

Fig. 7.21. Float-off processing using a multilayer resist.

Many multilayer schemes have now been reported exploiting any one or more of these advantages but they all have one common drawback, the extended processing, and therefore cost, which is required. Effort is presently being expended on ways to simplify the multilevel technique. Polysiloxanes, for example, have been used as imaging layers which incorporate the properties of the top two layers in a three-layer RIE system, thereby reducing the number of processing steps.[16] The development of simpler and easier multilayer systems will increase the use of processes, and may well become standard practice for difficult lithographic applications. This, as in other lithographic applications, will crucially depend on the development of new materials.

REFERENCES

1. DeForest, W. S., *Photoresist—Materials and Processes*, McGraw-Hill, New York, 1975.
2. Jagt, J. C. and Sevriens, A. P. G., *Polym. Eng. Sci.*, **20** (16) (1980) 1082.
3. Shiraishi, H., Taniguchi, Y., Horigami, S. and Nongaki, S., *Polym. Eng. Sci.*, **20** (16) (1980) 1054.
4. Thompson, L. F., Ballantyne, J. P. and Feit, E. D., *J. Vac. Sci. Technol.*, **12** (6) (1975) 1280.
5. Gong, B. M., Ye, Y. D., Gu, M. Q. and Zhang, Q. B., *J. Vac. Sci. Technol.*, **16** (6) (1979) 1980.
6. Lai, J. H. and Shepherd, L., *J. Appl. Polym. Sci.*, **20** (1976) 2367.
7. Ku, H. Y. and Scala, L. C., *J. Electrochem. Soc.*, **116** (7) (1969) 980.
8. Greeneich, J. S., *J. Appl. Phys.*, **45** (12) (1974) 5264.
9. See, for example, Billmeyer, F. W., *Textbook of Polymer Science*, Wiley-Interscience, New York, 1962.
10. Hiraoka, H. and Pacansky, J., *J. Vac. Sci. Technol.*, **19** (4) (1981) 1132.
11. Moran, J. M. and Taylor, G. N., *ibid.*, 1127.
12. Moreau, W. M., *Proc. Microcircuit Eng. Cambridge*, (1983) 321.
13. Batchelder, T. and Piatt, J., *Solid State Technol.*, August (1983) 211.
14. Taylor, G. N. and Wolf, T. M., *Polym. Eng. Sci.*, **20** (1980) 1086.
15. Pederson, L. A., *J. Electrochem. Soc.*, **129** (1) (1982) 205.
16. Hatzakis, M., Paraszczak, J. and Shaw, J. M., *Proc. Microcircuit Eng. (Lausanne)*, Sept. (1981) 396.
17. Thompson, L. F. and Kerwin, R. E. *Ann. Rev. Mat. Sci.*, **6** (1976) 267.
18. Pacansky, J. and Lyerla, J. R., *IBM J. Res. Develop.*, **23** (1979) 42.
19. Hieke, E. and Oldham, W. G., *Proc. Microcircuit Eng. Amsterdam*, (1980) 395.
20. MacDonald, S. A., Miller, R. D., Willson, C. G., Feinberg, G. M., Gleason, R. T., Halverson, R. M., MacIntyre, M. W. and Motsiff, W. T., *IBM Research Report*, RJ3624 (42420) 10 Jan. (1982).
21. Bowden, M. J., *J. Electrochem. Soc.*, **128** (5) (1981) 197C.

22. Bruning, J. H., *J. Vac. Sci. Technol.*, **16** (6) (1979) 1925.
23. Piwczyk, B. P. and Williams A. E., *Solid State Technol.*, June (1982) 74.
24. Helbert, J. N., Seese, P. A. and Gonzales, A. J., *SPIE*, **333** (1982) 24–30.
25. Hardy, C. J., Garside, J. R., Jones, P. L., Pickard, R. M. and Quayle, R. S., *Proc. Microcircuit Eng. Cambridge*, (1983) 83.
26. Kyser, D. and Viswanathan, N. S., *J. Vac. Sci. Technol.*, **12** (1975) 1305.
27. Alexander, P., Black, R. M., and Charlesby, A., *Proc. Royal Soc.*, **A232** (1955) 31.
28. Hatzakis, M., Ting, C. and Viswanathan, N., Fundamental aspects of E-beam exposure of polymer resists, *Electron Ion Beam Sci. and Technol. 6th Intl. Conf.*, San Francisco, 1974.
29. Hiraoka, H., *IBM, J. Res. Develop.*, **21** (1977) 121.
30. Alexander, P., Charlesby, A. and Ross, M., *Proc. Royal Soc.*, **A223** (1954) 392.
31. Moreau, W., Merrit, D., Moyer, W., Hatzakis, M., Johnson, D. and Pederson, L., *J. Vac. Sci. Technol.*, **16** (6) (1979) 1989.
32. Kakuchi, M., Sugawara, S., Murase, K. and Matsuyama, K. *J. Electrochem. Soc.* **124** (10) (1977) 1648.
33. Tada, T., *J. Electrochem. Soc.*, **130** (4) (1983) 912.
34. Roberts, E. D., *ACS Symp. Series*, **184** (1982) 1.
35. Thompson, L. F. and Bowden, M. J., *J. Electrochem. Soc.*, **120** (12) (1973) 1722.
36. Hiraoka, H. and Gutierrez, A. R., *J. Electrochem. Soc*, **126** (5) (1979) 860.
37. Bowden, M. J., Thompson, L. F., Fahrenholtz, S. R. and Doerries, E. M., *J. Electrochem. Soc.*, **128** (6) (1981) 1304.
38. Thompson, L. F., Ballantyne, J. P. and Feit, E. D. *J. Vac. Sci. Technol.*, **12** (6) (1975) 1280.
39. Tan, Z. C. H. and Georgia, S. S., *SPE Regional Technical Conf.*, Ellenville, New York, Nov. 1982.
40. Whipps, P. W., *Proc. Microcircuit Eng.*, Aachen, 1979, p. 118.
41. Thompson, L. F., Yau, L. and Doerries, E. M., *J. Electrochem. Soc.*, **126** (10) (1979) 1703.
42. Choong, H. S. and Kahn, F. J., *J. Vac. Sci. Technol.*, **19** (4) (1981) 1121.
43. Kamoshida, Y., Koshiba, M., Yoshimoto, H., Harita, Y. and Harada, K., *J. Vac. Sci. Technol.*, **B1** (4) (1983) 1156.
44. Imamura, S., Tamamura, T., Harada, K. and Sugawara, S., *J. Appl. Polym. Sci.*, **27** (1982) 937.
45. Ward, R., *J. Vac. Sci. Technol.*, **16** (6) (1979) 1830.
46. Shaw, J. M., Hatzakis, M., Paraszczak, J., Liutkus, J. and Babich, E., *SPE Regional Technical Conf.*, Ellenville, New York, Nov. 1982.
47. Hofer, D. C., Kaufman, F. B., Kramer, S. R. and Aviram, A., *Appl. Phys. Lett.*, **37** (3) (1980) 314.
48. Ito, H., Willson, C. G. and Frechet, J. M. J., *SPE Regional Technical Conf.*, Ellenville, New York, Nov. 1982.
49. Murase, K., Kakuchi, M. and Sugawara, S., *Intl. Conf. Microlithography*, Paris, June 1977.
50. Ryssel, H., Haberger, K. and Kranz, H., *J. Vac. Sci. Technol.*, **19** (4) (1981) 1358.

51. Yamaoka, T., Tsunoda, T. and Goto, Y., *Photog. Sci. Eng.*, **23** (1979) 196.
52. Taylor, G. N., *Solid State Tech.*, May 1980, 73.
53. Jain, K., Willson, C. G. and Lin, B. J., *IEEE Elec. Dev. Lett.*, **EDL-3** (3) (1982) 53.
54. Hartless, R. L. and Chandross, E. A., *J. Vac. Sci. Technol.*, **19** (4) (1981) 1333.
55. Wilkins, C. W., Reichmanis, E. and Chandross, E. A., *J. Vac. Sci. Technol.*, **127** (1980) 2510.
56. Imamura, S. and Sugawara, S., *Jap. J. Appl. Phys.*, **21** (5) (1982) 776.
57. Crivello, J. V., *SPE Regional Technical Conf.*, Ellenville, New York, Nov. 1982.
58. Reichmanis, E., Wilkins, C. W. and Chandross, E. A., *J. Vac. Sci. Technol.*, **19** (4) (1981) 1338.
59. Taylor, G. N., Wolf, T. M. and Moran, J. M., *J. Vac. Sci. Technol.*, **19** (4) (1981) 872.
60. Taylor, G. N., Wolf, T. M. and Goldrick, M. R., *J. Electrochem. Soc.*, **128** (2) (1981) 361.
61. Smith, J. N., Hughes, H. G., Keller, J. V., Goodner, W. R. and Wood, T. E., *Semiconductor International*, Dec. (1979) 41.
62. Tsuda, M., Oikawa, S., Kanai, W., Yokota, A., Hijikata, I., Uehara, A. and Nakane, H., *J. Vac. Sci. Technol.*, **19** (1981) 259.
63. Lin, B. J. and Chang, T. H. P., *J. Vac. Sci. Technol.*, **16** (6) (1979) 1669.
64. Moran, J. M. and Maydan, D., *ibid*, 1620.
65. Hatzakis, M., *J. Vac. Sci. Technol.*, **16** (6) (1979) 1984.

Chapter 8

PLASTICS FOR TELECOMMUNICATIONS

M. C. W. COTTRILL

Plessey Office Systems PLC, Beeston, Nottingham, UK

8.1. HISTORICAL REVIEW

The early history of telephony from the late 19th century onwards involved a limited range of insulating materials with rubber and ebonite most commonly employed by equipment designers. These materials were fashioned into components by many time-consuming machining operations.

The introduction of compression moulding opened up a whole new horizon for designers with the ability to produce a wide range of features in one operation. Components with flanges, holes and slots could now be produced by mass production moulding processes in thousands and hundreds of thousands using the new range of thermosetting moulding materials. Phenolics, urea formaldehyde and alkyd materials were employed for a whole range of components including fuse panels, coil formers, tag panels and, from 1929 onwards, telephone bodies and handsets.

The monopoly of thermosetting plastics was eroded during the 1940s by the introduction of the injection moulding process, which had been developed in the previous decade, involving the use of thermoplastic materials. The new process was much faster and a new range of materials with wide differences in properties now became available for consideration by designers. Materials such as cellulose acetate, polystyrene and acrylics became more and more widely used and manufacturing costs were significantly reduced.

During the 1950s the variety of thermoplastic materials increased and in particular nylon became more readily available. The incorporation of glass fibre reinforcement in nylon gave a material that had greater temperature resistance and which, compared to unfilled nylon, was stiff and creep-resistant. It could be moulded into thin sections and became the predominant material for the moulding of telecommunication components; millions of insulators, tag panels, bobbins and similar items were moulded in nylon during this period. The range of nylon types was expanded, but its relatively high moisture absorption and tendency to distortion led to its eventual displacement in many applications by other materials. In particular, polycarbonate with excellent electrical characteristics, low moisture absorption, good dimensional stability and low creep became widely employed during the 1960s.

The development and introduction of new materials continued through the 1970s with polyesters, polyphenylene oxide, polysulphones and indeed many other materials being utilised, but no one material can be said to have held a dominant position. More recently there has been a period in which each material has found its own particular place in the manufacture of equipment with a tailoring of materials for specific applications so that each material has become available in a wide range of grades. The choice available to equipment designers is now so wide that alternative materials can be chosen in many cases.

Connector devices give a good illustration of the range of materials chosen for one type of component. Influenced by detail specification differences or even the designers whim, connector mouldings are produced in polycarbonate, modified PPO, polyester, nylon and acetal.

To this point the usage of plastics within telecommunications has mainly involved electromechanical equipment, initially step-by-step Strowger-type, followed by Crossbar and more recently during the 1960s and 1970s Reed Electronic systems. During the 1980s, however, the silicon revolution has overtaken the telecommunications industry and, although electromechanical equipment is still being manufactured, all new development is concentrated on the introduction of silicon technology into all aspects of telecommunications equipment, main exchanges, private systems, and telephone and transmission equipment. The scope and the use of plastics have increased with each new generation of equipment.

Over the last few decades the number of plastic components has shown significant increase in both volume and variety but digital technology brings with it a countering factor in total material usage;

Fig. 8.1. Typical switching equipment—cabinets housing PCBs.

this factor is miniaturisation. The size of equipment of equivalent power has shown a marked downward trend; Strowger equipment providing 200 lines involved several cabinets occupying a room, similar Crossbar equipment would be half that size, the Reed Electronic system would be housed in one cabinet and digital equipment would involve one shelf, while the 'state of the art' development in VLSI would be one chip (Fig. 8.1).

8.2. PLASTICS USAGE

As already observed, the utilisation of plastics in telecommunications equipment has increased progressively with each new generation of equipment. The number of plastic components incorporated in the equipment has increased in both volume and variety, stimulated by economies that can be made and the range of features that can be moulded into one component. Material and tool developments have assisted this process by enabling the complexity of mouldings to increase.

A general survey will now be made of how plastics have been employed in the construction of the various categories of telecommunications equipment.

8.2.1. Exchange Equipment

The progress of plastics utilisation is well illustrated by exchange equipment where the early Strowger equipment incorporated insulating components fabricated from ebonite and wood- and resin-impregnated paper and textiles. Later developments introduced compression moulded items in thermosetting phenolics and alkyds and the later step-by-step equipment contained up to 20% of injection and compression moulded items.

Although the concept of the Crossbar system originated around the turn of the 20th century, it was not developed into a practical system until during the early 1920s. Its wider introduction was interrupted by the Second World War. The history of plastic materials in Crossbar systems begins in the main, therefore, with the era of compression moulding and it is probable that injection moulding with thermoplastics was already being introduced, certainly in the UK, before Crossbar systems became more widely manufactured and installed. This factor has contributed to the wider use of plastics in Crossbar equipment with the proportion of plastic material estimated at over twice that included in Strowger equipment.

The evolution of exchange equipment has continued through Reed Electronic systems to the latest digital electronic systems. This equipment, while having much greater switching power and a potential capability for providing a greatly enhanced service is, at the same time, much smaller. It contains few moving parts and the scope for plastic materials has increased so that we now find the proportion of organic material by volume to be 60–70% of the total material in the equipment.

The growth of the total organic material content of exchange systems in the form of plastic mouldings, wire and cable insulation and printed circuit board materials is illustrated in Fig. 8.2.

Typical examples of plastic components in exchange equipment are listed below; the items lower down the list relate more closely to the older electromechanical systems

(1) Connectors
(2) Printed circuit boards
(3) Card guides; handles and locking devices
(4) Integrated circuit packaging
(5) Console, VDU and other peripheral equipment housings
(6) Terminal blocks

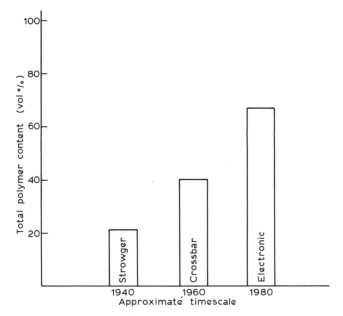

Fig. 8.2. Plastics in exchange equipment.

(7) Battery cases and cell components
(8) Wire and cable insulations
(9) Relay insulators and spacers
(10) Relay buffer blocks, coil formers
(11) Lever key components and insulators
(12) Jack strips
(13) Switchboard plug insulation
(14) Springset covers
(15) Rack covers
(16) Selector mechanism bank separators and insulators

The list is headed by connector mouldings which, although also relevant to electromechanical systems, now comprise the largest utilisation of plastic mouldings in new technology systems, second only in total quantity of organic material to printed circuit boards (PCBs). These applications will be examined in more detail later.

8.2.2. Customer Equipment
In this area of telecommunications, the most pronounced increase in the plastic content of equipment is found in telephone instruments and

the various items associated with private systems and the expanding area of office and business systems.

Although telephone instruments have existed for nearly 90 years, all the early versions were constructed of wood and metal and it can be said that plastics played little if any part in telephone development until 1929 when the first compression moulded sets were introduced. Subsequently very large numbers of black phenolic and coloured alkyd telephone body and handset mouldings were produced.

The next stage in the development of telephone instruments did not take place until the late 1950s when the now familiar injection moulded housings were introduced and it became possible to mould a set in thermoplastic on an 8 oz injection machine. Acrylic material was initially chosen for injection moulding, but was later superseded, because of its relatively low impact strength, by ABS which has maintained an almost exclusive place in the moulding of telephones to the present day (Fig. 8.3).

The design of telephone instruments has developed from the very early versions through the injection moulded type typified by the BPO

Fig. 8.3. Telephone instrument development.

Fig. 8.4. Typical electronic featurephone.

746 design in the UK (which contained a rotary dial—also eventually containing a number of engineering-type plastic mouldings), through the mechanical push-button versions to the latest all-electronic telephones. These now contain a minimum of metal components and not only have a moulded case and handset, but also contain at least one PCB with plastic encapsulated components, plastic keypads, plastic-based transducers, transmitters, receivers and tone callers (Fig. 8.4).

Telephone instruments now exist in great variety and are becoming more and more complex with an increasing number of features and functions, e.g. view-data phones incorporating television monitors for the display of information. Featurephones and small business systems also now have a very high plastic content. Their greater complexity mainly involves PCBs and the components mounted on them. They involve specially designed housings styled for functional reasons and to attract potential customers.

A novelty telephone market now also exists promoted, in the UK, by the liberalisation of the regulations covering telecommunications equipment. Here, in addition to antique reproductions, instruments are seen which incorporate moulded cartoon figures, such as Mickey Mouse and Snoopy, that are aimed at the younger members of the population.

Public telephones are often wall-mounted and protected by moulded acoustic hoods (Fig. 8.5).

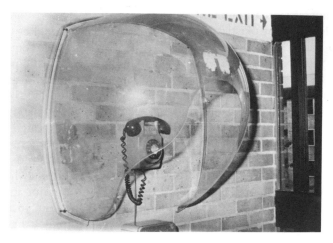

Fig. 8.5. Acrylic bubble acoustic hood for public telephone.

Customer equipment now involves the developing field of office and business systems. These systems may be self-contained, form part of a local area network or be based on a private digital exchange. They incorporate a whole range of peripheral equipment which, apart from the main operating consoles or work stations, also involve specialised telephones, word processors, facsimile machines, visual display units and data storage facilities.

A typical range of applications for plastics in customer equipment is listed below:

(1) Telephone bodies
(2) Handsets
(3) Console and equipment housings
(4) Keypads
(5) Printed circuit boards and components
(6) Connectors
(7) Switch mechanisms (hook switch)
(8) Transmitters/receivers
(9) Tone callers
(10) Ribbon cables
(11) Wires
(12) Call boxes
(13) Data storage media

8.2.3. Transmission Equipment

The development of plastics usage in transmission terminal equipment falls into much the same pattern as that for exchange equipment through to the present day, where examination of the latest equipment practices shows cabinets housing printed circuit boards with mounted components including many plastic encapsulated integrated circuits, connector devices, wires, cables and miscellaneous other plastic items.

Exchanges are linked by a network of cables which, although subject to continuing change through maintenance and augmentation, may in some places be half a century old. The earliest forms of transmission cable had rubber or gutta percha insulation, followed somewhat later by impregnated textile insulation and then in 1890 a major improvement was achieved by the introduction of paper insulation. In the UK it was not until the early 1950s that plastics began to play a significant role in the manufacture of transmission cables when polyethylene was introduced as the insulating material. In 1960 the same material, together with a wrapped aluminium foil moisture barrier, began to replace the hessian and pitch coated lead sheath which had in some form always been the standard sheathing system, although, due to initial difficulties with jointing, the change was not fully achieved until 1967.

In 1968 cellular polyethylene insulation fully filled with petroleum jelly was introduced. The new cable was smaller in diameter and lighter in weight making it easier to handle. It is now used in all new installations.

New developments involve optical fibre transmission cables which have a high plastic content and plastics also play their part in other transmission media—in communications satellites and submarine cables and at a more mundane level in the structure of telephone poles traditionally made of timber or, in some cases, steel.

8.3. APPLICATIONS

The selection of a plastic from the wide range available for a particular application will obviously be made from a knowledge of the performance required and, with differing emphasis, several of the properties listed below:

(1) Mouldability
(2) Impact strength

(3) Electrical characteristics
(4) Tensile strength
(5) Elongation
(6) Dimensional stability
(7) Temperature resistance
(8) Solvent resistance
(9) Creep
(10) Flammability
(11) Moisture absorption
(12) Hardness
(13) Abrasion resistance
(14) Thermal conductivity
(15) Frictional properties
(16) Optical properties
(17) Light fastness
(18) Chemical resistance
(19) Weathering resistance

The choice of material for particular applications will now be illustrated by examining a range of components.

8.3.1. Connectors

Connector devices of one form or another have always played a vital role in electrical equipment and the complex connectors now used, often in conjunction with printed circuit boards, form one of the major uses for organic materials in telecommunications equipment. Plugs and sockets and other early forms of connector devices were relatively large and cumbersome and originally moulded in phenolic or alkyd materials. Later, with the advent of injection moulding and thermoplastics, features of connector design such as the body, insulators, separators and lugs for snap-fit assembly could all be incorporated into one moulding.

The traditional thermosetting materials, filled phenolics and alkyds were rigid, had good heat and chemical resistance and, apart from a tendency to tracking, were excellent electrically. They were, however, difficult to adapt to automated assembly methods with snap-fit mouldings and push-fit contact designs which require a degree of material flexibility.

The requirements of plastics for connector applications include those listed below:

(1) Dimensional stability
(2) Toughness with flexibility
(3) Heat resistance
(4) Chemical/solvent resistance
(5) Electrical properties
(6) Non-flammability

The electrical properties must be excellent and the material must resist tracking. In an increasing number of cases, particularly in transmission applications, high-frequency properties are important. There must be no release of volatiles or migration of components of the material that could cause contact corrosion or in any way lead to a deterioration of contact performance.

Wave soldering operations can expose connector mouldings to temperatures in the region of 200°C for a few seconds and to organic solvents, normally of the halogenated hydrocarbon type, used for flux residue removal. Similar temperature and solvent exposures occur where vapour phase soldering is involved. High heat distortion temperatures and freedom from degradation at these temperatures, together with resistance to organic solvents, can therefore be essential requirements.

The trend for miniaturisation and increasing complexity of circuitry has already seen the standard for contact spacing fall from 0·2 to 0·1 in and a further reduction to 0·05 in may take place. The plastics employed for connectors must be easily moulded in intricate moulding tools with thin sections and close tolerances.

Low flammability is a necessity for most applications including those used in exchange equipment. Flame retardant grades are, therefore, normally employed and there is a move towards the use of plastics with inherently low flammability.

In the light of these requirements, and in the interests of increasing production rates and reducing costs, the move away from phenolic and alkyd thermosetting plastics has seen the selection by connector designers of a number of engineering thermoplastics. Materials currently in use include polycarbonate, modified polyphenylene oxide (PPO), polybutylene terephthalate (PBT) and polysulphones.

Polycarbonate and modified PPO are not suitable for applications requiring high temperature and solvent resistance. Polybutylene

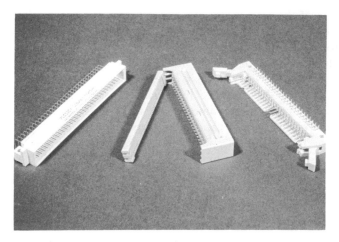

Fig. 8.6. Insulation displacement connectors—complex polyester mouldings.

terephthalate has a high heat distortion temperature and excellent solvent resistance, but requires careful moulding to avoid problems with dimensional stability.

Polysulphones have high heat distortion temperatures, but suffer from sensitivity to organic solvents. Although inherently non-flammable, they are expensive and therefore only used in limited connector applications.

Materials which may find favour in the future include polyphenylene sulphide (PPS) which has high temperature resistance and limited solvent sensitivity and polyether-ether ketone which has excellent temperature and solvent resistance, but, along with PPS, suffers from very high cost.

Insulation displacement connectors are gaining increasing favour and a typical example involving 0·1 in pitch contacts and intricate mouldings in PBT is illustrated in Fig. 8.6.

8.3.2. Printed Circuit Boards

Printed circuit boards are a relatively recent addition to the range of telecommunications components and although in simple form they have been used in telephone instruments since the middle 1960s, it was not until the advent of electronic switching systems that they became used in volume in exchange equipment. In the new technology digital-electronic systems printed circuit based materials constitute the largest proportion of the total material content and it has been estimated that the cost of the boards and associated costs amounts to some two thirds of the total cost of the equipment.

Double-sided, through hole plated boards are now the norm and multi-layer boards with four, six or eight layers (and even more in a limited number of cases) are increasingly used. Base materials are mainly epoxy resin impregnated glass fibre types although cellulose paper grades with both phenolic and epoxy resin impregnations are used for non-through hole plated boards. Flame retardant grades are required for most applications.

Dimensional stability is important for accurate registration of conductor patterns to drilled holes and to facilitate automatic insertion of components although the tolerances on hole locations have not proved to be as critical as once believed (Fig. 8.7).

Miniaturisation and higher component density, which are resulting in the use of surface mounted leadless components, produce problems of heat dissipation and mismatch of thermal expansion of component package and base board that may produce fatigue stress in soldered joints and potential failure.

The trends in interconnection technology in general and PCBs in particular are towards higher component density, increasing use of surface-mounted components and base material developments including the incorporation of laminated metal cores or the substitution of

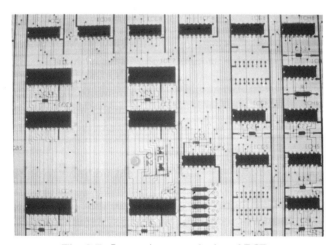

Fig. 8.7. Increasing complexity of PCBs.

the glass fibre by aromatic polyamide fibre (aimed at improved electrical, mechanical and thermal performances).

A much more detailed review of printed circuit board materials is given in Chapter 9.

8.3.3. Housings

The properties required of materials for equipment housings include most importantly, surface finish, impact strength, low flammability and low cost.

Since the early 1960s acrylonitrile–butadiene–styrene (ABS) has gained and retained a predominant position for the moulding of telephone bodies and handsets; its impact strength, good surface appearance, ease of moulding and reasonable cost have led to its almost exclusive use. Although extensive programmes of evaluation of alternative materials have been conducted, only ASA and rigid PVC have been identified as potential alternatives. More recent pressures on cost could, however, result in the use of polystyrene against reduced specification requirements, particularly in respect of impact strength. Polycarbonate has been used for certain vandal-proof telephone installations (Fig. 8.8).

Mains voltages are now appearing in certain special feature telephones (for example, in those which incorporate CRT monitors) and in these cases flame retardant grades are specified.

Larger housings for the keyboards, consoles and other peripheral equipment associated with newer technology systems, including key

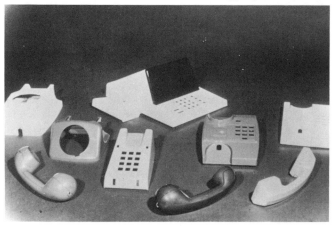

Fig. 8.8. Telephone instrument mouldings in ABS.

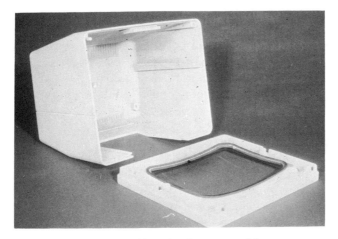

Fig. 8.9. VDU housing in structural foam.

systems, have seen the employment of foamed moulding materials. The increasingly competitive market for this type of equipment requires that it should be attractively styled as well as functionally sound in order to attract and retain potential customers. It is in this area that structural foams offer a number of advantages. The foam structure is achieved by the addition of blowing agents which either react or evaporate to produce gaseous products at moulding temperatures so that much lower processing pressures are required to fill the mould (Fig. 8.9).

The use of structural foam moulding allows more freedom in design. A minimum wall thickness of about 6 mm is required, but thick sections, bosses and strengthening ribs can be incorporated without the risk of sink marks on the exposed outer surfaces.

Structural foam mouldings offer scope in design and economy in cost particularly with respect to tooling cost; their main features are:

(1) rigid and strong—6 mm wall section
(2) thick sections, ribs, bosses, etc., produce no sink marks in outer surfaces
(3) freedom from moulded in stress and dimensional stability facilitate easy assembly
(4) moulded cases can support functional components—power supplies, CRTs etc.—eliminating chassis and subframes

(5) reduced tooling costs—lower pressures allow use of unhardened steel or aluminium tools

(6) the cost of presses is less than for equipment involved in the injection moulding of components of equivalent size

(7) structural foam mouldings have a quality feel and their use avoids the cheap and nasty image sometimes associated with plastics

Paint finishing is normally required to achieve a good surface finish and appearance. Due to their cellular structure, the surfaces of foam mouldings are rough and also affected by a swirling pattern caused by entrapment of the blowing agent gases between the moulding and the cavity walls. The swirling pattern may be minimised by attention to tool temperature and speed of injection, but so far has not been eliminated. The use of textured or spatter paint coatings is normal to reduce surface preparation and hide minor defects. In some cases if high gloss is required, filling, flatting and multiple paint coats may be necessary. Polyurethane enamels are normally employed.

Sandwich moulding offers an alternative which avoids the need for finishing foam mouldings, but so far has found little application for telecommunications as the very high tool and machine costs, together with initial process development expenditure, necessitate very long runs for economic justification. In this process a shot of unfoamed material such as polyethylene is followed by the main shot of foamed plastic to produce foamed mouldings with a dense non-porous skin and good surface finish.

EMI/RFI shielding, intrinsic with metal cabinets, is being increasingly specified for plastic housings and this is generally achieved by spraying the interior surface of the mouldings with a conductive paint. An alternative technique involves vacuum deposition of a metal coating.

Structural foam materials currently being used include modified PPO, ABS, polycarbonate, polypropylene and polystyrene, all in flame retardant grades.

Rigid integral skin structural foams are produced in polyurethane materials by the reaction injection moulding (RIM) process. Components of the polymer are metered via a mixing chamber to give a measured short shot in a tool cavity where the polymerisation and the foaming action, which expands the material to fill the cavity, take place. At the cavity-moulding interface the bubbles formed by the blowing reaction are collapsed by the cool surface and a solid skin is produced. The density of the foam core is controlled by means of the amount of blowing agent incorporated and the tool temperature (normally 55–60°C) so that plastic with the desired rigidity, density and skin thickness can be produced.

Features of polyurethane structural foams produced by the RIM process are:

(1) Latitude in design is afforded by the possible use of moulding weights from a few grams to many kilograms together with less restriction on flow length to wall thickness ratios than for thermoplastic foam.
(2) In contrast to thermoplastic foams, the tool surface controls the finish of the moulding so that excellent surface finishes can be achieved.
(3) Polyurethane foams are discoloured on exposure to UV radiation so that it is generally still necessary to finish with a polyurethane enamel as for thermoplastic foams.
(4) The mouldings can be sawn and drilled, but if screwed fixings are required, it is generally better to mould the necessary holes to establish skin with greater strength than the foam core.

The RIM process involves relatively low tool pressures and temperatures so that low-cost tooling such as cast epoxy may be used for short runs. This is very attractive for large items in small quantities and is especially suited to the production of prototype housings. For example, certain business equipment housings were moulded as prototypes in polyurethane, but larger quantities were injection moulded in foamed modified PPO.

8.3.4. Circuit Components
Plastics in the form of epoxide moulding compounds have now almost completely dominated the packaging of integrated circuits which, when mounted on printed circuit boards, form the functional core of the latest technology equipment.

In the early days of electronic systems, their specifications were naturally influenced by the performance traditionally expected of the older electromechanical systems with 20 to 30 year lifetimes typically being guaranteed. In this context plastic packaged active as well as passive components were found to have reliability shortcomings. Extensive research and development has taken place aimed at improving the performance of such devices and moving their reliability closer to that of the metal and ceramic hermetically sealed packages initially utilised.

A number of factors have enabled the economies in cost possible with the use of plastic packaged devices to be realised. The development of new grades with freedom from the ionisable impurities that could result in corrosion problems when transported to the chip surface by water permeation, the incorporation of corrosion inhibitors, the careful selection of plastic material (to avoid thermal expansion mismatches with resulting stresses on the chips) and the reduction in life expectancy of current equipment due to the rapid rate of development and outdating of equipment associated with the new silicon technology has now resulted in plastic encapsulated devices dominating the component field.

8.3.5. Card Guides, Handles and Locking Devices

Printed circuit boards with their mounted components form the major feature of all electronic switching systems. Examination of equipment cabinets generally reveals them mounted in rows located by card guides. They are retained in many cases by special latching devices and their removal is often facilitated by handles attached to the front edge of the boards. All these items are normally moulded in plastic (Fig. 8.10).

The design of card guides involves a number of features with

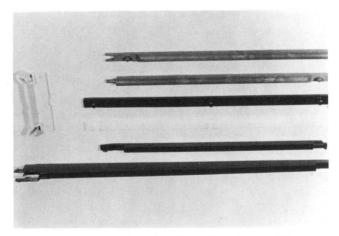

Fig. 8.10. Typical card guide and handle mouldings.

conflicting property requirements. The card guides which are relatively long and slender in section must support the weight of the assembled printed circuit boards and not creep under the sustained load. They must also have low friction surfaces to facilitate the sliding into position of the boards. Fixing of the guides into the main card cage often involves a snap fit feature requiring flexibility and fatigue resistance particularly if repositioning of the guides is envisaged. Inevitably with these requirements some compromise in material properties is necessary. Materials that have been chosen for this application include nylon, polycarbonate and modified PPO in both unfilled and glass filled grades.

8.3.6. Switches

Cradle, gravity or hook switches, as the enabling switches on telephone instruments are variously known, have not escaped the plastic invasion. Various designs have seen the mechanism replaced, all or in part, by plastic components. In one low-cost telephone design the gravity switch consists of a series of slides, pivots and levers moulded in nylon, polycarbonate and polyacetal mounted in the ABS body moulding, the choice of material for separate parts of the switch being determined in each case by the main function of the part concerned.

8.3.7. Dials and Keypads

One of the most familiar parts of a telecommunications system to the user both in the home and in business has been the rotary dial. This was originally almost entirely constructed of metal components, but later developments incorporated an increasing number of plastic components, particularly in the form of gears in acetal. Eventually almost entirely plastic dials with smooth and less noisy operation were in use.

More recently, push-button key switch units were introduced. Early forms of this mechanism involved contact springs mounted in complex base mouldings, operated by moulded buttons and plunger assemblies.

The latest types of key switching unit are constructed almost completely from plastics. The spring action is provided by a silicon rubber mat with conductive inserts which are pressed directly on to gold-plated contacts on a printed circuit board by two-colour buttons moulded in ABS. The main body of the unit with the button guides is moulded in glass filled modified PPO. The relatively high expenditure associated with the two-colour moulding of the operating buttons has led to the development of single shot moulded buttons onto which the legend is printed. Inks and printing processes have been developed

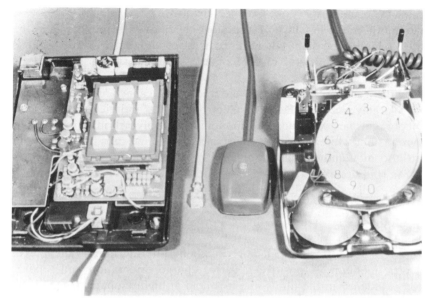

Fig. 8.11. Inside the telephone: left, new electronic; right, older electromechanical.

that produce sufficiently durable lettering and figuring on the buttons
(Fig. 8.11).

Other types of key unit incorporate Hall effect switches while touch
types using inductive or capacitative properties have also been employed for certain keyboards.

A device which is finding some favour for telephone and keyboard
applications is the so-called membrane keypad. This is a basically
simple unit made from polyester or polycarbonate film in which the
contact members, conductive leads and the button pattern are applied
by screen printing methods. Two opposing contact patterns are separated by a layer of unprinted polyester film having holes corresponding
to the key pattern. Pressure on the appropriate printed key area on the
face of the pad brings the two contact surfaces together with resulting
electrical connection. This device has attractions on both cost and
technical grounds, but has found user resistance because of the 'dead-
ness' of feel and lack of positive response to key operation. Various
developments are in hand involving the inclusion of spring washers to
produce 'oil-can' effects and raised 'button' areas to simulate the more
conventional button-operated units.

8.3.8. Transducers

Yet another area in which plastics have taken over from metal is in the latest telephone designs where on cost, space and performance grounds traditional transmitters, receivers and bells have been replaced by specialised transducers.

In the case of transmitters and receivers, carbon granule devices have largely been displaced by electromagnetic and electrical devices. Diaphragms are produced from polyester, polycarbonate and polystyrene films while the main bodies or housings and caps for the devices are also moulded in plastic for both economy and user safety.

The traditional metal bells in telephones have largely been displaced by tone callers, typically involving a piezo-ceramic diaphragm retained in a plastic case.

Another type of device under development for the conversion of speech into electrical signals is the so-called PVDF microphone which employs specially processed polyvinylidene fluoride film as the piezoelectric medium. A positive–negative charge orientation and separation is induced in the film while it is held at high temperature in an oil bath. The charge orientation is retained on cooling back to normal temperatures. Subsequent displacements of diaphragms made from this film, metallised on both sides and typically around 0·030 in thick, produce changes in the potential difference across the film and these can be amplified into electrical signals of appropriate strength.

8.3.9. Data Storage Media

The new telecommunications technology, like the computer technology on which it is based, has an ever increasing requirement for data storage and this particularly applies to business and office systems. Magnetic media provide the means of storing data and exist in three major forms: tape, rigid disc and flexible disc. In all three cases a plastic substrate is coated with a film of magnetic material such as iron oxide.

Magnetic tape is based on a polyester substrate and is contained on plastic reels housed in a plastic cassette.

Flexible or floppy discs were developed early in the 1970s, initially in 8 in disc form and later on as the $5\frac{1}{4}$ in mini-disc. Flexible disc storage media have become standard for the industry. The substrate in this case is a polyester–polyethylene terephthalate disc, again coated with ferrous oxide. The discs tend to be supplied in envelopes in cardboard containers, but plastic containers are being considered to provide better protection for the rather vulnerable discs. A further development is the 3–4 in diameter 'micro' disc, the actual size varying from manufacturer to manufacturer. In this case the discs are contained in rigid polypropylene cases.

The third type of medium, the rigid disc, makes less use of plastics, the disc substrate being polished aluminium. Rigid discs were introduced in a 14 in diameter version contained in a plastic cartridge, but the trend has been for miniaturisation and a number of Winchester disc devices are now in use. The plastic used for the cartridges is typically polycarbonate.

8.3.10. Batteries

Emergency power supplies involve standby rechargeable batteries traditionally of the lead–acid accumulator type with vulcanised rubber cases and cellulosic separators.

Most of the development work taking place in battery design is aimed at traction applications where the requirement is for high energy density, low weight and low cost. The scope for plastics has widened to include the structure of the battery cells and electrodes as well as the case. Batteries incorporating a much higher proportion of plastics could find their way into telecommunications applications because of better efficiency, longer operational life and lower cost.

The requirements of a battery case include chemical stability in contact with the electrolyte, enough strength to support the weight of the electrodes and electrolyte without deforming and possibly transparency or translucency to allow viewing of the electrolyte level. Plastic cases have advantages in all these respects.

Materials chosen for battery cases include polypropylene, ABS, SAN and high-impact polystyrene. Nylons can be used for alkaline-type batteries but are unstable if exposed to acid electrolytes.

Plastics such as polypropylene are now being considered for electrode supports while polyethylene, polypropylene and, in certain cases, polyamides, polystyrene and PVC may be used for electrode spacers. In microporous form, produced by polymer sintering techniques, most of these materials are used for battery separators which are semi-permeable membranes acting as metal ion barriers.

Cellulose polymers may be added to electrolytes as gelling agents while solid polymer electrolytes (inert polymer matrices with mobile charged groupings attached) are possible future developments.

Batteries are required as a back-up to solar panels used as power sources in communications satellites and this is one area where (via low weight and high energy density designs) plastics are being employed in space applications.

8.3.11. Wire and Cable

Before the Second World War, the standard insulation for wire used in telecommunications in the UK was waxed acetylated cotton, the conductor being tinned copper wire, while for tropical use the conductors were oleo-resinous enamelled copper wire. Jumper wire had braided wool insulation.

PVC-coated wire insulation was introduced in Germany during the war period, but this suffered from stress cracking and it was not until around 1950 that a hard non-cracking PVC insulation was developed.

Over the next two decades extruded PVC with a lacquered, lapped rayon secondary coating for protection against soldering temperatures and for identification was the standard wire insulation. The jumper wire over this period was PVC-coated tinned copper wire with a cellulose lacquered rayon braid secondary coating containing a non-flammability additive.

A relatively stable era ensued in which better grades of PVC extrusion were developed and the printing of colour spirals on the wire for identification purposes was introduced. In the late 1960s bar marking was introduced. The identification inks were applied to the PVC wire covering during the extrusion process, the heat of the PVC being used to dry the ink. This allowed much faster speeds of extrusion to be obtained with benefits on cost and capacity.

Later in the 1970s nylon-coated PVC jumper wire was used in the UK, the nylon coating giving toughness and cut-through resistance, but at the same time introducing a flammability problem. This was rapidly overcome, however, by the introduction of a single shot extrusion of PVC crosslinked by irradiation.

More recently the trend in equipment wiring has been for the use of small diameter, 0·25 and 0·4 mm wires, subject to sharp bends over closely spaced tags. Over the last 4 to 5 years attention has turned to the requirements of automatic wiring and cut, strip and wrap applications. The demand has been for tight tolerances to be maintained on tough thin wall insulation while producing controlled adhesion of insulation to conductor wire to ensure satisfactory stripping. As a result, polyvinylidene fluoride, PTFE and heat-sealed, wrapped polyester wire insulation have been introduced, the last giving high cut-through resistance for the smallest gauges. Crosslinked versions of polyvinylidene fluoride are being developed for still higher cut-through resistance.

Prior to the Second World War, equipment cabling was sheathed with painted wool braid and this continued in the UK into the early 1950s when extruded PVC-coated wires were in use. Subsequently, flexible PVC sheathing was introduced.

The increasing use of insulation displacement connectors involves both wire and ribbon cable and, although these still employ PVC insulation, alternatives under investigation include polyvinylidene fluoride, modified PPO and polyesters.

Wire and cable joints are often protected by the use of irradiated PVC heat-shrink sleeving.

The flammability of wire and cable insulation has been an issue for much development and discussion since although PVC is inherently a low flammability material at normal temperatures it burns readily at higher temperatures and in a fire situation can offer an excellent path for flame propagation. In the few fires that have occurred in telecommunications switch equipment, the major damage has been caused by corrosion due to acidic hydrogen chloride gas evolved from the burning PVC rather than by the fire itself.

Alternative plastics have been evaluated for wire insulation, but have not so far been introduced. Emphasis has been placed on fire prevention and in certain cases on the introduction of special fire barriers in exchange cabling runs.

8.3.12. Optical Fibre Cable

Optical fibre technology is playing an increasingly important role in telecommunications and will be indispensable in the systems of the future to ensure that the capacity of transmission networks matches the capability of the switching equipment to generate, process and control vast amounts of information. Optical fibre cables have exceptional capacity and rates for the transmission of voice, video and data signals (Fig. 8.12).

Plastics play a significant part in the structure and manufacture of optical fibre cables. The individual fused silica fibres need a protective coating which has conflicting requirements. It needs to be soft enough to cushion the fibre and protect it from surface damage and to minimise microbending losses while at the same time being hard

Fig. 8.12. Optical fibre cable.

enough to withstand impact and abrasion during subsequent manufacturing operations, so that transmission characteristics are not impaired
and survival of the fibre for the life of the cable is ensured. The
requirements can be met by employing a double coating with a silicone
primary layer and a secondary harder layer of UV-cured acrylic
polymer. Typically, a 200 μm diameter fibre would be coated with
50 μm of the primary layer and 30–40 μm of the secondary layer. An
alternative secondary coating system can involve a loosely fitted polypropylene tube. The coated optical fibres are laid up to form cable with
polyester yarn filler members and high-strength aromatic polyamide
members to protect the fibres from the strains occurring during cable
insulation. Polyester tape wraps are used to maintain the registration
of the cable components during manufacture. An overlapped
polyethylene–aluminium foil film is used to prevent the ingress of
water vapour which can cause fibre breakage due to a corrosion
mechanism. The whole cable is substantially sheathed in polyethylene.

Plastic optical fibres in the form of polystyrene or acrylic fibres are
already in use for up to 100 m links where data rates are relatively low.
Plastic optical fibres offer lower cost and greater flexibility, but are
subject to much greater attenuation losses and development is continuing which could lead to the much wider use of plastic-cored fibres.

The foregoing applications all have relevance to new technology telecommunications systems. Nevertheless, step-by-step and Crossbar equipment are still being manufactured and examples of plastics components in the older type of equipment will now be given.

8.3.13. Terminal Blocks
These are essentially blocks of an insulating medium containing an array of conducting tags and they serve to provide a means for permanent connections to the terminal posts or tags.

Originally fabricated from ebonite or wood and later compression moulded in phenolic or alkyd, they are now injection moulded in glass-filled polycarbonate or glass-filled nylon.

8.3.14. DSU Plugs and Sockets
This older generation type of connector was used for inter-rack connection on the Crossbar systems installed by the British Post Office during the 1970s. They involved a complex base moulding originally made in phenolic or alkyd, but later moulded in glass-filled polycarbonate or nylon. The mouldings for both plug and socket members were identical and in their ultimate form were injection moulded in unfilled polycarbonate. The design incorporated contact separators and insulators in a one-piece moulding.

8.3.15. Jack Strip Mouldings
Jack strips are associated with manual switchboards and the insulating members involved again demonstrate the development of moulding

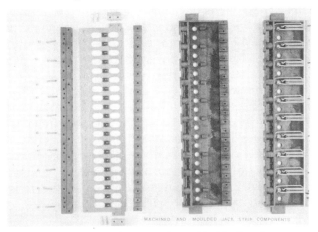

Fig. 8.13. Machined and moulded jack strips.

technology beginning with the machining of the parts in ebonite, developing to compression moulding in phenolic and ultimately to injection moulding. The latest designs exhibit a diversity of form partly made possible by material and process advances. Glass-filled nylon was used for the mouldings for individual jacks, but later multiple designs were moulded in glass-filled polycarbonate because of distortion problems with nylon (Fig. 8.13).

8.3.16. Key Switch Insulators

Miniature lever key insulators show similar development; initially compression moulded in phenolic and later glass-filled epoxy material, they were then injection moulded in glass-filled nylon. At least one manufacturer made a material change to glass-filled polycarbonate, but stress cracking problems due to the use of cleaning solvents to remove flux residues led to a further change to glass-filled modified PPO—the material choice being constrained by the necessity to continue with existing tooling.

8.3.17. Springset Separators

The insulating members that separate the contact springs in relay and other springset applications were originally machined from phenolic resin bonded sheet material. The clamping screws were sheathed with ebonite tubes to prevent shorting between springs. The application involves continuously applied clamping stresses and, in the case of the British Telecom 3000 type Relay, the separators are injection moulded in glass-filled polycarbonate which will withstand the continuous load without excessive creep. Identification of insulators with marginally different thicknesses is achieved by colour coding—using the facility offered by plastics for variously coloured grades.

8.3.18. Relay Buffer Blocks

The British Telecom 3000 type Relay again illustrates an evolution involving the use of plastics. Originally compression moulded in alkyd, it was ultimately produced in glass-filled polycarbonate which gave the necessary combination of dimensional stability and low creep while being mouldable with greater precision at lower cost.

8.3.19. Springset Lifting Combs

Various materials have been used for relay lifting combs; acetal and nylon, both glass-filled and unfilled, and polycarbonate. The general material requirements in this case are for dimensional stability, freedom from frictional wear with no debris to contaminate the contacts, mouldability to give accurately dimensioned parts and, of course, good electrical insulation properties.

In one particular design which incorporated an adjusting screw, a self-locking action for the screw was also required; nylon 11 was found to be satisfactory for this application.

8.3.20. Springset Covers

Protective covers are fitted to springsets in certain cases. If visibility is not important then modified PPO or polyethylene may be employed; where visibility is required clear polycarbonate, CAB or acrylic material can be used.

8.3.21. Equipment Rack Covers

PVC has been used to seal racks containing Strowger equipment against adverse environments as, for example, in the case of exchange equipment installed in a geothermal area. This has been developed more generally as a means of excluding dust and in certain cases, where more rigid covers were required, in the form of hinged doors and lids, clear polycarbonate or acrylic plastic has been employed.

8.4. MATERIALS AND PROBLEMS

A wide range of the available plastics is used in telecommunications equipment and in this section the most common materials are listed. In certain cases their shortcomings are illustrated by examples of problems encountered.

8.4.1. ABS

Acrylonitrile–butadiene–styrene, a modified polystyrene, is the material most widely used for the injection moulding of telephone housings. It exhibits good mouldability with good surface finish and impact strength, and is available in flame retardant grades. It has been used for other applications such as connectors, but its poor solvent and temperature resistances have led to its displacement from applications demanding these properties.

8.4.2. Acetal

Acetal or polyoxymethylene, a crystalline material, exhibits good wear resistance and a low coefficient of friction. It has been used for bearings and moving parts such as gear wheels and spindles in rotary dials. Acetal is easily processed and has good resistance to organic solvents, but below average resistance to high temperatures and flammability led to its displacement.

In one application involving a dial spindle, moisture absorption caused swelling and seizing of the dial mechanism. The mechanism had to be redesigned with attention to clearances.

8.4.3. DAP

Diallyl phthalate, alkyd materials once widely used along with phenolics in the era of compression moulding, provide excellent electrical characteristics, rigidity and chemical resistance. They are used instead of phenolics in spite of higher cost when justified by their much better electrical performance in humid environments and their pale natural colour which allows a much better range of coloured mouldings to be achieved.

8.4.4. Modified PPO

Styrene-modified polyphenylene oxide is used in both filled and unfilled grades. It can be produced in low flammability versions by the incorporation of flame retardant additives. It has good mouldability in all but the most complex tools and is fairly widely employed. In some cases it has taken over from polycarbonate.

8.4.5. Nylon

The nylons (or polyamides) provide a range of materials with excellent mouldability and with the inclusion of glass filling having moderate temperature and creep resistances. Although once widely used for relay coil formers, connector mouldings and insulators in general, their high moisture absorption and tendency for distortion resulted in their eventual replacement by alternative materials.

8.4.6. Polycarbonate

This material appeared at one time to be the answer to many moulding problems with excellent electrical stability and creep characteristics, together with good resistance to elevated temperatures and excellent impact strength. It was found, however, to be very susceptible to degradation during moulding if the moulding powder was not thoroughly dried, and it suffered from sensitivity to solvents and stress cracking, particularly under constantly applied localised stress. These problems were evident in the case of a miniature lever key insulator. The design featured an interference fit retention of the contact springs which produced a constant stress in the moulding. The addition of a solvent cleaning process to this situation, to remove flux residues following the soldering of wires to the contact springs, produced an epidemic of moulding disintegration and the change to an alternative material.

8.4.7. PVC
Polyvinyl chloride is widely used in extruded form for wire insulation and cable sheathing. Although alternatives have been sought, it remains the predominant material for this application. Irradiated, crosslinked grades are now in use where greater toughness and cut-through resistance are required.

As a moulding material, PVC was considered as an alternative to ABS for telephone housings and its applications included selector mechanism bank separators (Fig. 8.14). In selector mechanism bank separators it suffered from ageing, plasticiser migration and, with particular relevance to the selector application, was chemically degraded by the zinc content of finely divided brass or nickel silver wear debris. The problem was minimised by attention to material grade since the mechanism was found to relate to the lubricant and stabiliser contents of the plastic.

Fig. 8.14. PVC bank separator degradation.

Table 8.1

Polymer Applications

Application	ABS	Acetal	Acrylic	CAB	DAP (alkyd)	Epoxide	GRP	Nylon	Phenolic	Polycarbonate	PPO-modified	Polyester, PBT, etc.	Polysulphones	PPS	PVC	Polyethylene	Polypropylene	Polystyrene	Polyurethane	Structural foams	Polyvinylidene fluoride	Silicones	PTFE
Connectors	●	●			●	●			●	●	●	●	●	●	●								
Printed circuit boards						●			●														
Integrated circuit packaging						●																●	
Card guides, etc.								●		●	●												
Telephone bodies and handsets	●								●														
Housing-consoles, VDUs, etc.	●																	●	●	●			
Wire and cable equipment									●						●						●		●
Transmission cable																●							
Optical fibre cable				●				●								●	●						
Transducers										●	●	●									●		
Data storage media										●		●											
Batteries	●															●	●	●	●				
Keypads	●	●								●												●	
Hook switch mechanisms		●							●			●	●										
Terminal blocks					●					●	●	●											
Springset insulators										●	●												
Lever key components		●			●					●	●	●											
Coil formers		●								●	●		●										
Relay buffer blocks					●	●			●			●	●										
Selector mechanism insulators																			●				
Rotary dials		●							●														
Springset covers				●	●					●	●	●							●				
Rack covers				●								●							●				
Jack strips									●	●	●												
Switchboard plug insulation					●				●														
Telephone poles							●																

8.4.8. PBT

Polybutylene terephthalate offers a high heat distortion temperature and excellent solvent resistance, especially in the glass-filled grades, and it is becoming widely used in applications such as connector mouldings where these properties may be required. The moulding conditions require careful control since, to avoid dimensional instability, high melt temperatures must be employed leading to the possibility of thermal degradation.

8.4.9. Polysulphones

These are a range of materials with high heat distortion temperatures in the region of 200°C and they are inherently non-flammable. Their sensitivity to organic solvents and high cost, however, limit their use to certain special applications.

8.4.10. PPS

Polyphenylene sulphide is a newer material that has good moulding characteristics and may be considered for telecommunications applications. It has a higher heat distortion temperature than PBT, is inherently non-flammable and is resistant to most solvents.

8.4.11. PEEK

Polyether-ether ketone is among the new materials that offer very high temperature resistance, chemical resistance and stability but, because of its currently very high cost, has yet to be used in other than extremely specialised applications.

A summary of polymer usage in telecommunications applications is given in Table 8.1. This gives a general idea of materials that have been used; it is not necessarily comprehensive and the information is not exclusive nor does it differentiate between grades or the use of filled or unfilled plastics.

8.5. FUTURE TRENDS

The new dimension of information handling power that has become available is creating a revolution in telecommunications during the last two decades of the 20th Century. The telecommunications system is becoming a medium for the communication of total information involving voice, video and data signals.

Utilisation of the new technology has resulted in developments such as integrated digital exchanges with associated business information systems and, potentially, the automated office.

The trend for miniaturisation is countered by the considerably increased proportion of plastics in telecommunications equipment and the overall consumption of plastics by the industry has increased significantly.

Automatic assembly processes will not only add to the scope for telecommunications equipment but will also make their own special demand on plastics, including heat and solvent resistances.

The market for telecommunications equipment is becoming increasingly competitive and attractive appearance is essential so that plastic housings offer the possibility of producing attractive shapes in a range of colours at relatively low cost.

While nylon, polycarbonate, modified PPO and particularly ABS continue to have wide application, polyesters are finding increasing favour where heat and solvent resistances and mouldability to fine tolerances are required. Polysulphones have little application and newer materials such as PPS and polyether-ether ketone may find a place in special applications where the requirement for stability, and heat and solvent resistances justifies their high cost.

Modification and alloying in order to tailor existing materials for particular blends of useful properties means that care must be exercised in choosing the correct grade of material for each application.

Much attention is being paid to tool design with computerised techniques being used to optimise gating and material flow within the mould cavity in order to reduce moulded-in stress and to maximise component strength, dimensional precision and stability.

Good mouldability, surface finish, stability, heat and solvent resistances, low flammability and always reasonable cost are among the main requirements of equipment designers.

Plastics and polymeric materials in general already constitute a major proportion of the material content of the telecommunications equipment now being designed and manufactured. The opportunities for plastics in telecommunications equipment are great and the use of plastics by the industry will remain at a high and increasing level for many years to come.

ACKNOWLEDGEMENTS

The author wishes to acknowledge the helpful discussions with colleagues in Plessey, other telecommunications companies and British Telecom that have contributed to this chapter. He also thanks the Manufacturing Executive, Standard Systems, Plessey Office Systems PLC for permission to undertake this work.

Chapter 9

PLASTICS FOR PRINTED WIRING SUBSTRATES

R. HURDITCH*

PPG Industries, Inc., Pittsburgh,
Pennsylvania, USA

9.1. INTRODUCTION

Printed circuits were developed in the USA during World War II primarily to reduce weight and space and increase reliability in military equipment. This technology replaced discrete wiring of electrical components which was previously accomplished by the use of terminal boards and point-to-point wiring and soldering.

In recent years the term 'printed wiring' has been used synonymously with printed circuit and is technically more correct, since a circuit includes components and devices.

Printed wiring is a pattern of electrical conducting paths reproduced onto an insulating medium designed to interconnect electronic components and devices.

A variety of reprographic and printing techniques is used to fabricate the pattern of conducting traces and pads (usually copper) which form the printed wiring. Some of these methods will be described later on in this chapter. This review is principally concerned with the polymeric materials which are used to fabricate the insulating medium which supports the printed wiring. However, a basic knowledge of the types and applications of printed wiring is necessary in order to understand the factors which influence the choice of materials used.

277

9.2. TYPES AND END USES OF PRINTED WIRING

There are two basic types of printed wiring: rigid and flexible. Rigid printed wiring is usually constructed from a reinforced thermosetting plastic substrate commonly referred to as a printed wiring board or printed circuit board. Rigid printed wiring is used whenever a rigid plane or configuration is required for interconnection (Fig. 9.1). Flexible printed wiring is made from a flexible base material which is used to interconnect components assembled in a non-planar arrangement such as folding around bends. Flexible printed wiring is often employed where space is limited (Fig. 9.2).

Rigid printed wiring is used far more extensively than flexible, and in the USA accounted for about 88% of the total area produced (approximately 250 million square feet) in 1983.

Both rigid and flexible printed wiring are further classified according to structure of which there are three basic types, as illustrated in Fig. 9.3:

(1) Single sided—conductors on one surface of the insulating substrate or base.
(2) Double sided—conductor on both sides of the substrate.

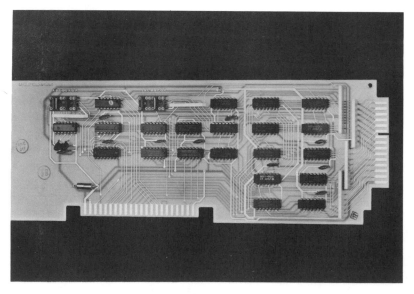

Fig. 9.1. Rigid printed wiring boards for computer applications.

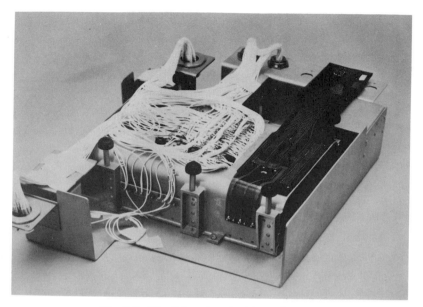

Fig. 9.2. Flexible printed wiring substrate.

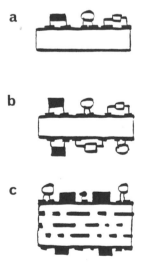

Fig. 9.3. Schematic illustration of the three basic types of printed wiring. (a) Single sided; (b) double sided; (c) multilayer.

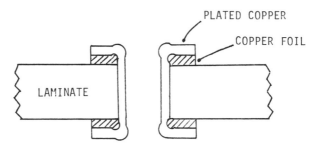

Fig. 9.4. Schematic illustration of a plated through hole.

(3) Multilayer—three or more conductor layers separated by insulating medium.

The individual conducting planes in double-sided and multilayer substrates are usually interconnected by holes, the walls of which are plated with copper, as illustrated in Fig. 9.4. These are usually referred to as plated through holes.

There are four major electronic equipment end uses or markets for printed wiring: computers or data processing, communications, military and consumer. These accounted for about 85% of the total US market for printed wiring in 1983, estimated to be valued at \$3·6 billion.[1] The remainder comprises institutional, industrial and instrumentation markets. Multilayer and double-sided printed wiring are the types most commonly used in computers, communications and related equipment whereas single sided is used primarily for consumer electronics.

Advanced computers, in particular, require printed wiring with a very high interconnection density to cope with the large number of input and output terminals on large-scale integrated circuits (LSI and VLSI), such as memory and microprocessor chips. An impressive example of state-of-the-art use of multilayer wiring is provided by the IBM 3081 board which has external dimensions of 60×70 cm $\times$ 4·6 mm thick and contains 20 discrete conducting layers in all.[2] Much of the stimulus for improvements in printed wiring manufacturing techniques and materials derives from integrated circuit development for use in advanced electronic equipment where performance frequently outweighs cost considerations.

However, another important contributing factor to the developments in printed wiring also derives from consumer electronics where reduced manufacturing costs and cheaper materials are of prime consideration. These considerations will be discussed in later sections of this chapter.

For those who desire to learn more about printed wiring markets and technology trends, a number of informative reports can be purchased from sources given under Ref. 3.

9.3. LAMINATES FOR RIGID PRINTED WIRING

Nearly all (rigid) printed wiring boards in current use are laminates made by bonding together several sheets of fibrous reinforcing material which is surface clad with copper foil on one or both sides. The fibrous sheets are usually in the form of woven fabric, mat or paper and made of glass, often an electronic grade (E-glass), cellulose or synthetic polymers. These are frequently referred to as base materials. The binding agents are thermosetting resins including epoxy, phenolic and polyester.

9.3.1. Manufacturing Process

The manufacture of a laminate involves three basic steps, namely treating, laying up and pressing. In the first operation the base material is passed through a dip tank containing a solution of the resin. After saturation, excess resin is removed by squeezing between rollers and the material is dried in an oven to a partially cured state or B-stage. The semicured material is known as a prepreg and is normally dry and tack-free. The exact ratio of resin to base material, and the degree and uniformity of cure are all carefully controlled in the treatment operation. For the lay-up or build-up operation, copper foil is laid against a polished steel press-pan, and a number of sheets of prepreg are laid on top to achieve the desired thickness for the type of board. For two-sided boards a second sheet of copper foil is placed on top of the stack which is capped with another press-pan. The final operation takes place in a laminating press which may be capable of laminating 100 or more sheets 1×2 m $\times 1 \cdot 5$ mm thick. The hydraulic presses can be operated at high pressure (2000–3000 psi).

Each press platen is uniformly heated and the pressure–temperature cure cycle is carefully controlled to minimise non-uniformities, undercure, void formation and excessive stress build-up.

Table 9.1
NEMA Copper-Clad Laminates

NEMA grade	Resin	Reinforce-ment	Description	Use
XXP XXXPC	Phenolic Phenolic	Paper Paper	Hot punching grade Room-temperature punching grade	Inexpensive laminates mainly used in low-cost consumer items, e.g. calculators, watches
FR2	Phenolic	Paper	Flame retardant equivalent of XXXP/XXXPC	Used extensively in consumer items, e.g. radios, smoke detectors; low-cost substitute for FR4 in single-sided wiring
CEM1	Epoxy	Paper + glass	Punchable epoxy, properties between XXXPC and FR4	
G10	Epoxy	Glass cloth	General-purpose glass base; excellent electrical properties and water resistant	Widely used in more demanding applications, e.g. computers, telecommunications; relatively expensive (approx. $2 \times$ FR2)
FR4	Epoxy	Glass cloth	Similar to G10, but flame retardant	
FR5	Epoxy	Glass cloth	Similar to FR4 but more thermally stable	Generally specified for military/aerospace applications

9.3.2. Types and End Uses

Most of the laminates widely used for printed wiring boards are classified by the National Electrical Manufacturers Association or NEMA (ASTM standards are identical). NEMA standards are based on the combination of electrical, mechanical, thermal and chemical properties which determine the suitability of a laminate for a particular end use.

In the NEMA classification system, paper (cellulose) reinforced grades are designated with the letter X, glass fabric grades with the letter G and flame retardant grades with the letters FR. There is generally a flame retardant equivalent for each of the X and G types which may differ only slightly in specifications. In addition to NEMA standards there are, of course, military (MIL) specifications which generally demand a higher level of performance. However, most commercial glass-reinforced copper-clad laminates qualify under both NEMA and MIL standards. Detailed descriptions and comparisons of these specifications can be obtained from the IPC.[4] The most widely used NEMA grade printed wiring board laminates are listed in Table 9.1 which also briefly describes their composition, basic features and end uses.

The epoxy glass-fabric grades possess superior electrical, mechanical and chemical properties compared with the phenolic paper grades; the latter grades, however, are about one-half the cost (approximately US $1·00 per sq. ft.). Laminates which classify under the flame retardant grade FR4 account for over 50% of all of the types produced in the USA and are available in thicknesses ranging from 0·8 to 3 mm (0·031 to 0·125 in).

Although NEMA specifications provide a basic guide to properties, the actual behaviour of laminates of the same grade during the processing steps required to produce a printed wiring board may differ considerably. Important factors that influence the detailed properties (apart from the resin composition, which will be described in subsequent sections) are the thickness and weave style of the glass fabric and the bonding or coupling agent used to promote adhesion. For more information on this aspect of laminate fabrication the reader is referred to the publications given in Ref. 5.

9.4. PRINTED WIRING FABRICATION

The following brief description of printed wiring fabrication will serve to acquaint the reader with some of the basic steps which are referred to in subsequent sections dealing with the properties of substrate materials.

The fabrication of printed wiring is primarily carried out by replication of an image of the circuit pattern onto a copper-clad substrate which is etched to form the wiring. This is referred to as a subtractive process since copper is removed. In recent years, an alternative technique called additive processing has gained in popularity, particularly in Japan. In additive processing, copper is electro-deposited imagewise onto a bare substrate. The substrate is activated to accept the copper by prior treatment with a special coating.

In addition to the chemical-based processes there are also a number of mechanical methods based on direct wiring.

The first stage in the manufacture of printed wiring involves preparation of the artwork depicting the circuit pattern. This is then reduced to a photomask which is the same size as the final substrate.

The substrate processing operation depends upon whether the type of wiring to be produced is single sided, double sided or multilayer. Essentially identical processes are used for both rigid and flexible substrates.

9.4.1. Single Sided: Subtractive

Following a preliminary rigorous cleaning and degreasing process, the copper-clad substrate is coated with a negative photoresist (which is rendered insoluble by UV irradiation), and then exposed to near-UV radiation through the mask. This reproduces the circuit image onto the copper which is then developed by dissolving the unexposed photoresist. An alloy of tin and lead (60:40) is electroplated over the exposed circuit pattern. The alloy increases the solderability of the copper and serves as an etch resist. Residual photoresist is removed (stripped) with a solvent and the exposed copper is etched with an acid etchant. Finally, holes are drilled or punched and the substrate rigorously cleaned and degreased prior to component assembly and soldering.

9.4.2. Double Sided: Subtractive

Processing a double-sided substrate involves the additional procedure of forming interconnecting holes (plated through holes). In other respects the process is identical to that described in Section 9.4.1.

Plated through hole fabrication is the initial step and involves drilling or punching the holes in the virgin copper-clad board at the locations required for interconnecting circuitry between the two wiring planes. This is followed by cleaning, and electroless copper is then deposited over the whole substrate including the hole walls. The copper plate wall thickness may then be built up by electroplating (see Fig. 9.4). After a further cleaning operation photoresist is applied and the process continued as described for a single-sided board.

9.4.3. Multilayer: Subtractive

The preparation of a multilayer board involves laminating together *in precise registration* individual two-sided boards. Each plane of circuitry is separated by layers of prepreg prior to the lamination process.

No wiring pattern is formed on the top and bottom layers of the stack since a continuous conducting plane is required for the formation of plated through holes which is the next step following lamination. Finally, the wiring patterns are formed on the top and bottom planes.

9.4.4. Additive Process

The additive process involves no etching, and starts with an unclad substrate that is coated with an epoxy-based microporous adhesive material. This provides an anchor for the copper, which is deposited by an electroless process. Through hole plating and wiring pattern plating are carried out simultaneously. Additive processing is applicble to most substrate materials, and to single, double or multilayer wiring fabrication.

9.4.5. Semiadditive Process

This is essentially a hybrid method in which copper is first deposited onto the whole of either one or two sides of an unclad substrate using the additive process. The wiring pattern is defined in the conventional manner using photoresist and copper is electroplated to build up the thickness of the wiring pattern. Photoresist is stripped and the background electroless copper is removed by a quick etch.

9.5. PROPERTIES AND TESTING OF MATERIALS FOR PRINTED WIRING

The general physical and chemical properties which are required of a printed wiring substrate are largely determined by the following factors:

(1) *Electrical*
    (a) Excellent insulating properties retained under prolonged ex-

posure to humid environments and high circuit operating
temperatures.
(b) Low dielectric loss.
(c) Uniform electrical properties.
(2) *Mechanical/thermomechanical*
(a) Drillable or punchable.
(b) Dimensional changes induced by heat and/or chemical treatments should be reversible, reproducible, uniform and small.
(c) Rigid substrates should have minimal warp and twist.
(d) Resistant to the effects of mechanical vibration.
(e) Resistant to thermal shock induced by soldering (260–275°C
for 5 s in wave soldering, 215–245°C for 20–60 s in vapour
phase reflow soldering).
(3) *Chemical*
(a) Resistant to acids, alkalis and oxidants used in copper cleaning, etching and plating.
(b) Resistant to organic solvents used in cleaning/vapour degreasing (e.g., methylene chloride, trichloroethane).
(c) Strong adhesion to copper retained under adverse environmental and processing conditions.

Some of the desired properties such as the electrical performance are
relatively easy to measure and interpret. However, thermomechanical
characteristics, such as dimensional stability and warp and twist, depend upon a number of interacting complex physical and chemical
properties including differential thermal expansion, swelling and stresses built in during cure or lamination. They are therefore more difficult
to predict or quantify in a precise manner from simple measurements
of basic properties.

This is especially true when considering a laminate, which is anisotropic in nature, i.e. the reinforcement acts in the XY plane but not
significantly in the Z axis, for which the thermal expansion and
mechanical properties are largely determined by the resin.

Printed wiring manufacturers rely heavily on a large number of tests
that are primarily designed to detect failures under conditions which
reproduce processing or extreme use. These procedures are covered
extensively by publications of studies instituted by the IPC.[4] In the
following, the basic properties and test methods generally used to
characterise plastic materials for printed wiring applications are briefly
described. Recommended ASTM testing standards are listed in Table

Table 9.2
Standard Test Methods

Method	*ASTM test no.*
Electrical resistance	D257
Dielectric breakdown voltage, electric strength	D149
Dissipation factor and dielectric constant	D150
Arc resistance	D495
Flexural properties of plastics	D790
Impact strength	D256
Shear strength	D732
Punchability	D617
Warpage of sheet plastics	D1181
Thermal expansion	D696
Deflection temperature under load	D648
Thermal stress	E2/125
Chemical resistance of plastics	D543
Water absorption of plastics	D570

Note: flammability is normally tested according to Underwriters Laboratory Procedure UL 94.

9.2. Additional test methods and standards for printed wiring laminates are described in MIL-P-13949 and IPC-TM-650. Properties for a number of polymeric and other electronic materials can be found in the Appendices A1 and A2.

9.5.1. Test Conditioning

It is customary to define the environment in which tests are performed on electrical and electronic grade materials. The conditioning designations are:

(1) Condition A—as-received; no special conditioning
(2) Condition C—humidity conditioning
(3) Condition D—immersion conditioning in distilled water
(4) Condition E—Temperature conditioning

In designating specific test conditioning the letter (C to E) is followed by three numbers separated by obliques. The first represents the duration in hours, the second is the conditioning temperature in degrees centigrade and the third is the relative humidity, assuming humidity is controlled. For example, C-96/35/90 which is a common

conditioning for electrical resistance measurements, indicates humidity
conditioning for 96 h at 35°C and 90% relative humidity.

9.5.2. Electrical Properties

9.5.2.1. Surface and Volume Resistivity

Resistivity measurements are obtained from resistance measurements
using electrode arrangements of defined geometries and are a measure
of the vitally important insulating characteristics of printed wiring
materials. One of the commonly used electrode arrangements is the
circular three electrode pattern illustrated in Fig. 9.5 (ASTM D257).
For surface measurements the bottom electrode is the guard conductor
which intercepts stray (bulk) currents. Resistance measurements are
usually made with an applied potential of 500 V. Surface resistivity, ρ_s,
is calculated from the relationship

$$\rho_s = \frac{R'P}{D}$$

where R' is the surface resistance, P is the perimeter of the guarded
electrode and D is the distance between the inner and guard electrodes. The units are expressed as Ω/square. Typical values for polymers lie in the range 10^7–10^{10} Ω/square. Surface resistivity decreases
with increase in humidity or water absorption. The magnitude of this
effect can be greatly increased if ionic impurities are present on the
surface which become conducting on exposure to water. It is common
practice to carry out measurements after humidity conditioning
(Method C).

Volume resistivity is obtained from volume resistance using the
arrangement shown in Fig. 9.5 but with electrode 2 as the guard and
with the DC potential applied between 1 and 3. The resistivity is
calculated from the relationship

$$\rho = \frac{RA}{d}$$

where R is the volume resistance, A is the effective area of electrode 1
and d is the specimen thickness. Typical values for electrical grade
polymers are of the order of 10^{15} Ω cm at room temperature. Volume
resistivity usually exhibits an exponential decrease with increase in
temperature and is also significantly reduced by water absorption.

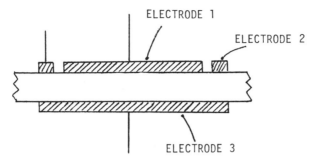

Fig. 9.5. Electrode arrangement for surface and volume resistivity measurement.

Resistivity values for a selection of polymeric materials are given in Appendix A1.

9.5.2.2. Dielectric Constant

The dielectric constant is the ratio of the capacitance formed by two parallel plates with a material between them to the capacitance with air as the dielectric. A uniform dielectric constant is particularly important for substrates used in high-frequency and microwave electronic circuits.

A low dielectric constant is particularly desirable for two-sided or multilayer boards used for high-speed computer circuits or other high-speed applications where distributed capacitance is important. For the glass fabric resin systems such as FR4 this can be reduced by increasing the resin-to-glass ratio since the E-glass used has a dielectric constant of 6·1 compared with approximately 3·5 for a typical epoxy resin. Figure 9.6 shows the effect of resin content on dielectric constant.

The dielectric constants of most polymers are relatively low compared with inorganic materials and lie in the range 2 to 4 (at 1 MHz) with a few materials outside this range (see Appendix A1).

9.5.2.3. Dissipation Factor and Loss Factor

The dissipation factor is the ratio of the parallel reactance to parallel resistance, i.e. the tangent of the loss angle (tan δ). The loss factor is the product of tan δ and the dielectric constant and is a measure of the total loss of power occurring in the insulating material. Dissipation factors for most plastics tend to decrease with increasing frequency and

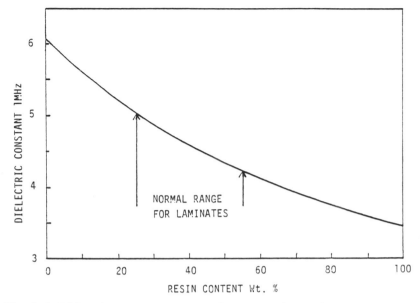

Fig. 9.6. Dielectric constant versus resin content for a glass-epoxy system. Epoxy resin, $K = 3 \cdot 5$; E-glass, $K = 6 \cdot 1$.

increase with increasing temperature. For printed wiring applications values are usually reported at 1 MHz, and at room temperature are in the range 3×10^{-2} to 3×10^{-4}. A low dissipation factor is also important for printed wiring used in high-frequency electronic circuits.

9.5.2.4. Electric Strength

Electric strength (also referred to as dielectric strength) is the field strength in volts per unit thickness at which electrical breakdown occurs. It is normally measured in accordance with ASTM D149 by applying a 60 Hz voltage through the thickness of an oil-immersed sample, as shown in Fig. 9.7(a). The applied voltage is increased at a uniform rate smoothly or step-wise. The magnitude of the electric strength depends upon the thickness of material, the form and size of the electrodes, the frequency and waveform of the applied voltage and the material properties. Hence, it is unwise to compare dielectric strength data for different materials unless all test conditions are standardised.

Fig. 9.7. Electrode arrangements for (a) dielectric strength (field perpendicular) and (b) dielectric breakdown measurements (field parallel).

9.5.2.5. Dielectric Breakdown

Dielectric breakdown is a measurement of breakdown voltage which is made parallel rather than perpendicular to the plane of a sample, as illustrated in Fig. 9.7(b). In the standard test (D149) the electrodes are taper pins inserted into the sample on 1 in centres. The voltage is applied at a controlled rate of increase with the sample immersed in oil. Dielectric strength depends on both surface and bulk characteristics of a material and is, of course, reduced by increases in humidity, water absorption and ionic or other conducting contaminants. Typical values for polymeric material are in the range 10–50 kV.

9.5.3. Mechanical and Thermomechanical Properties

9.5.3.1. Flexural Strength and Modulus

Flexural strength is the force per unit area required to break a material in bending. Flexural modulus is the ratio of the elastic limit of stress to corresponding strain and relates to the stiffness of a material under load. Load–deformation measurements are obtained by applying a force to the centre of a sample in the form of a beam or bar which is supported at both ends as illustrated in Fig. 9.8. The flexural modulus (FM) is calculated from the slope m of the steepest initial portion of the

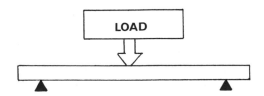

Fig. 9.8. Schematic illustration of flexural strength test.

load–deformation curve by the equation

$$FM = \frac{L^3 m}{4bd^3}$$

where L = span, b = width and d = depth of the beam under load.

Modulus values vary from $0 \cdot 3 \times 10^6$ psi for thermoplastics to 5×10^6 psi for reinforced laminates.

9.5.3.2. Deflection Temperature and Glass Transition Temperature

Deflection temperature, more commonly known as heat distortion temperature, is a measure of the thermal stability of a plastic material. A widely used procedure is ASTM D648, according to which the sample arrangement is similar to that used for the measurement of flexural strength (Fig. 9.8). A standard load (usually 264 psi) is applied to the centre of the sample which is supported with a span of 4 in. The temperature at which the sample deflects $0 \cdot 010$ in upon heating at a rate of 2°C/min is reported as the deflection temperature. The deflection temperature of a polymeric material may approximate to the glass transition temperature (T_g) as determined for example by thermomechanical analysis (see Chapter 2). The agreement is likely to be closer for thermoplastic materials which soften reversibly above T_g. For a thermoset, however, the thermomechanical behaviour can be complicated by the degree of cure. Thus undercured materials may become more rigid on heating and exhibit irreversible heat deflection behaviour. In the case of a laminate, additional complications may arise due to stress relief.

In general, above the heat distortion or glass transition temperature most fully cured thermosetting polymers lose their rigidity and hence T_g is an indication of the upper limit of service temperature for rigid printed wiring materials. In laminates the reinforcing effect may extend temperature serviceability above T_g, where dimensional stability is not critical as in the case of single-sided printed wiring boards.

Glass transition temperatures for a selection of polymeric materials are listed in Appendix A1.

9.5.3.3. Thermal Expansion Coefficient and Dimensional Stability

The coefficient of thermal expansion is a measure of the dimensional changes which occur on heating or cooling a sample. The values normally measured are linear and reported at one particular temperature which, in the ASTM D696 standard method, is 55°C (130°F). The coefficient is the change in length per unit length per degree in temperature.

In general, polymers have larger coefficients of expansion than most of the other materials used in electronic circuits (see Appendix A2). In addition, the thermal expansion coefficients of polymers tend to increase above the glass transition temperature. This effect is illustrated in Fig. 9.9. Since printed wiring boards are subject to many thermal cycles, in order to minimise stresses it is desirable to minimise the difference in thermal expansion between the copper foil and the laminate. This can be reasonably achieved in the case of an epoxy–glass laminate since the expansion of copper (16·5 ppm/°C) lies between that of E-glass (2·8 ppm/°C) and a typical epoxy resin (40–50 ppm/°C) at room temperature. Thus values for the room-temperature expansion coefficient of an FR4 laminate are typically in the range 10–15 ppm/°C in the XY plane. However, in the Z axis, the value is generally in the range 40–50 ppm/°C, which is close to that of the pure epoxy resin. Figure 9.10 shows the Z axis thermal expansion coefficient of the FR4 laminate as a function of temperature. Here it can be seen that there is a dramatic increase in the expansion in the Z axis through the glass transition temperature. This

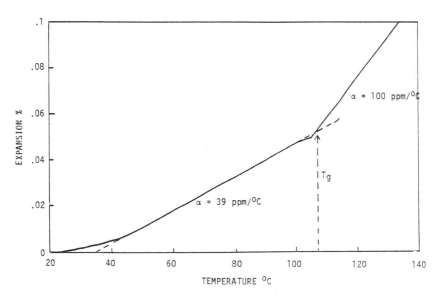

Fig. 9.9. Typical expansion curve for a cured DGEBA epoxy resin.

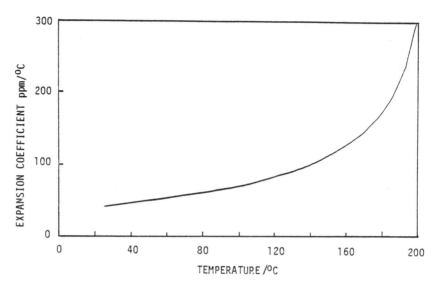

Fig. 9.10. Z axis expansion coefficient versus temperature for a typical multi-layer epoxy laminate.

phenomenon results from the increase in the thermal expansion coefficient of the pure epoxy resin above T_g (Fig. 9.9), coupled with the transfer of volume expansion to the Z axis due to the in-plane confining action of the glass fabric.

This behaviour can result in severe problems due to cracking of plated through holes for laminates exposed to temperatures close to or higher than T_g. In addition, the copper foils in the inner layers of multilayer boards can undergo similar cracking. Thus excellent dimensional stability in the XY plane is achieved at the cost of low Z axis stability. This problem has precipitated a considerable interest in the development of higher T_g epoxy resins and the use of alternative polymers such as polyimides. Some of these materials will be reviewed in the following section.

9.5.4. Water Absorption

High water absorption can result in pronounced reduction in the insulating resistance, dielectric breakdown, dissipation factor and mechanical strength. Water absorption is dependent upon the chemical nature of the polymer as well as upon the presence of voids, which is generally greater in composite materials such as laminates. In the fabrication of printed wiring substrates considerable care is taken to minimise void formation. In the case of a laminate this requires thorough impregnation of the reinforcing material with the resin and complete removal of all volatile materials.

9.6. THERMOSETS FOR LAMINATES

Nearly all laminates suitable for printed wiring applications are based on thermosets. Compared with the commonly available thermoplastics, thermosets possess improved temperature performance and chemical stability due to the crosslinking which occurs on curing to produce a three-dimensional rigidised chemical structure. The most important polymers used for the fabrication of printed wiring laminates are described in the following.

9.6.1. Epoxy Resins

Epoxy resins possess a combination of desirable properties which have resulted in their widespread use for the fabrication of laminates. The most important of these are: high adhesive strength; high mechanical strength; good electrical insulation properties; low shrinkage; excellent chemical resistance and thermal stability; ease of cure; and ready modification to optimise properties by blending and by use of different curing agents.

Currently over 90% of thermosets used for printing wiring laminates in the USA are epoxy-based. The resins which are employed almost exclusively are those based on the diglycidyl ether of bisphenol A (DGEBA) and the tetrabromo derivative which is used to provide flame retardant properties. These compounds are represented by the following formulae:

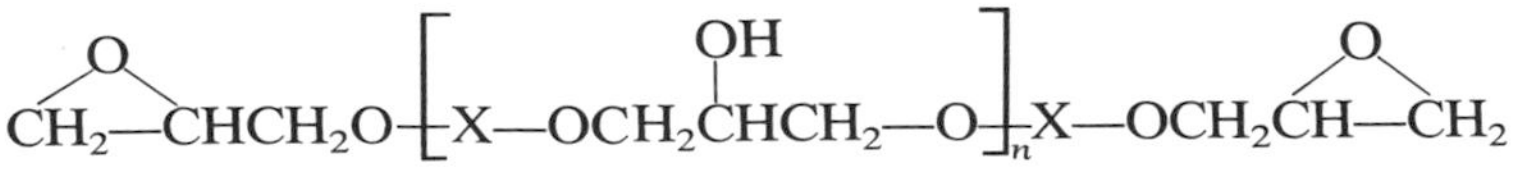

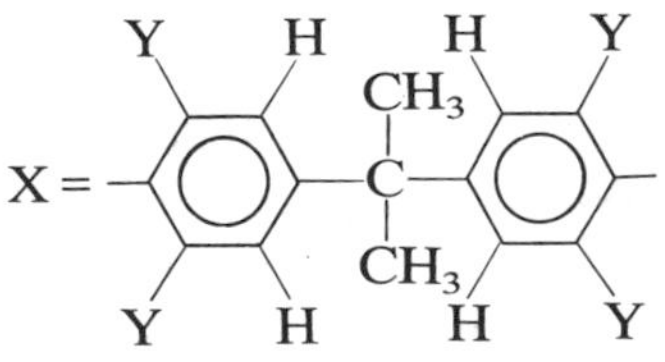

$$\text{CH}_2\text{—CHCH}_2\text{O} \left[\text{X—OCH}_2\overset{\text{OH}}{\underset{}{\text{CHCH}_2}}\text{—O} \right]_n \text{X—OCH}_2\text{CH—CH}_2$$

$$\text{X} = \text{—} \underset{\underset{\text{Y} \quad \text{H}}{}}{\overset{\overset{\text{Y} \quad \text{H}}{}}{\bigcirc}} \text{—} \underset{\text{CH}_3}{\overset{\text{CH}_3}{\text{C}}} \text{—} \underset{\underset{\text{H} \quad \text{Y}}{}}{\overset{\overset{\text{H} \quad \text{Y}}{}}{\bigcirc}} \text{—}$$

where $Y = H$ or Br. A description of the chemistry of these and related epoxy compounds is given in Chapter 4.

The choice of DGEBA resins for use in laminates is based on the higher thermal and chemical stabilities which characterise aromatic systems, coupled with the relatively low cost of raw materials. The resins are used with n values in the range 0–2. A typical formulation would have an average molecular weight of 900 ($n = 2$ for the non-brominated DGEBA) and a solids content of 20% in a solution of acetone or cellusolve.

Aromatic polyamines are commonly used as curing agents, particularly, metaphenylenediamine (MPDA), 4,4'-methylenedianiline (MDA) and diaminodiphenylsulphone (DADS), the formulae of which are as follows

MPDA

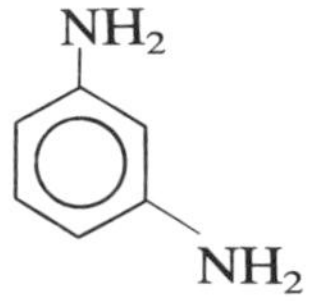

Melting point, 60°C

MDA H_2N—⟨○⟩—CH_2—⟨○⟩—NH_2 Melting point, 170–180°C

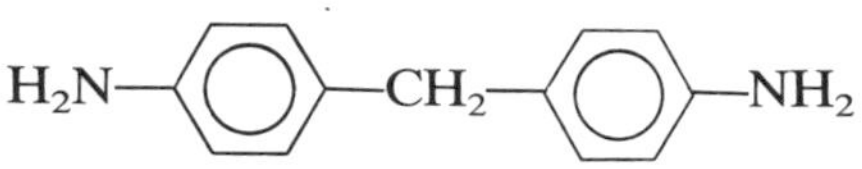

DADS H_2N—⟨○⟩—$\overset{O}{\underset{O}{S}}$—⟨○⟩—$NH_2$ Melting point, 207–209°C

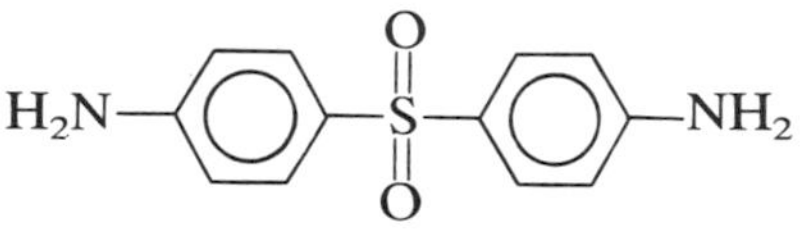

The curing of these compounds with DGEBA is normally carried out in two stages, typically 2 h at 125°C followed by 2 h at 160–200°C.

For FR4 grade laminates and prepregs, the most commonly used curing agent is dicyandiamide (DICY)

$$\underset{\displaystyle H_2N—\overset{\textstyle NH_2}{\overset{|}{C}}=N—C\equiv N}{}$$ Melting point, 208°C

In the presence of a suitable accelerator, typically benzyldimethylamine (BDMA), at concentrations of 3–3·5% by weight, laminated DGEBA and DICY cure in 1–1·5 h at 160°C. This fast and relatively low temperature cure greatly facilitates multilayer fabrication where

Table 9.3
Typical Properties of Laminating Epoxy Resin Systems Based on DGEBA

| Property | Curing agent[a] | | | | MPDA + 67% glass fabric | Brominated DGEBA + MDA |
	DICY	MDA	DADS	MPDA		
Flexural strength (psi×10⁻³)	18	17	16	19	80–90	17
Flexural modulus (psi×10⁻⁶)	0·44	0·39	0·42	0·46	3·8–4·2	0·41
Tensile strength (psi×10⁻³)	9	9·5	9	8	50–58	8·2
Deflection temperature[b] (°C)(264 psi)	127	155	175	155		157
Dielectric constant (25°C)	3·9	3·7	3·9	3·8	5·1	3·4
Dissipation factor (1 kHz, 25°C)	0·012	0·017	0·013	0·018	0·009	
Dissipation factor (1 MHz, 25°C)	0·021	0·035	0·027	0·035	0·019	0·033
Volume resistivity (Ω cm ×10⁻¹⁵)	10	2	0·1	10	0·2	10
Water absorption (%)(24 h)	0·03	0·04	0·035	0·03	0·05–0·07	0·04

[a] Stoichiometric amounts used.
[b] Optimised cure cycle.

higher temperatures are less desirable because of increased dimensional changes. Furthermore, the B-stage material has excellent long-
term stability.

Typical physical and electrical properties of DGEBA cured with
polyamines are shown in Table 9.3.

It can be seen that the electrical and mechanical properties are very
similar. The main differences are in the heat distortion temperatures
which increase in the order DICY < MPDA < MDA < DADS.

Recently there has been considerable effort devoted to the development of epoxy systems which possess all the desirable characteristics of
an FR4 resin but exhibit improved dimensional stability, particularly
for multilayer fabrication. The approach generally pursued is to aim
for a system which has its glass transition temperature above the cure
temperature, thus ensuring that extensive thermal expansion does not
occur during fabrication.

Several proprietary laminating epoxy resin systems have been reported,[6] for which glass transition temperatures in excess of 170°C are
claimed but with processing conditions similar to those of the
DGEBA/DICY system.

9.6.2. Polyimides

Polyimides were developed during the 1960s and early 1970s in
response to the demands of the aerospace industry for high temperature performance polymers, as matrix materials for laminates and
other composites. Most aromatic/heterocyclic polymer systems that
have a small number of oxidisable C–H bonds per molecule exhibit
excellent oxidative stability, but tend to be intractible and extremely
difficult to process. In the case of polyimides, however, this limitation
has been largely overcome. Some of the materials now in use for
structural applications can withstand continuous exposures in air to
temperatures above 300°C (approximately 600°F).

The characteristic structural feature of a polyimide is the presence of
the phthalimide grouping in the repeat unit, which is represented by
the formula

These units may be linked through alkyl or aryl groups to form the main polymer chain, the latter generally giving higher temperature performance.

The most general method for the preparation of high temperature polyimides is based on the reaction between an aromatic diamine and an aromatic dianhydride or tetracarboxylic acid.[7] This proceeds via the formation of the intermediate polyamic acid which in further heating undergoes a ring closure condensation reaction leading to imide formation. The reaction is as follows

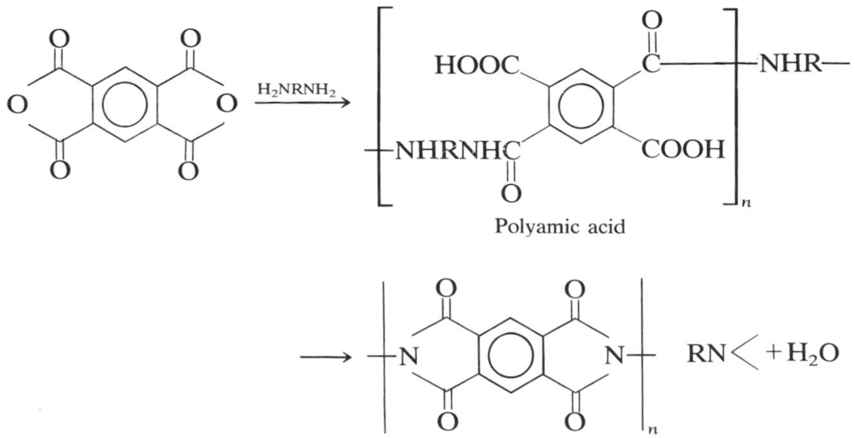

Polyamic acid

The fact that water is eliminated during the final cure to produce a highly condensed polymer with rapidly increasing viscosity (low flow) results in a relatively high concentration of voids, removal of which requires heating above T_g (typically >290°C) under pressure for several hours. For this reason, the condensation polyimides have not been successfully applied to the fabrication of printed wiring and laminates, although polyimide films prepared by this route are used in other electronic applications (see Chapter 10), including flexible film substrates (as described in Section 9.8).

In the late 1960s another class of polyimides was developed by Rhone Poulenc in which the anhydride or acid is replaced by a bismaleimide and the resulting polyimide is formed without elimination of water. The reaction illustrated schematically below proceeds via amine addition about the double bonds of bismaleimide to form a linear prepolymer.

In the presence of excess bismaleimide and a free radical initiator, self-addition can occur at the maleic double bond leading to crosslinking which produces a polymer of high thermal stability.

Laminates based on Rhone Poulenc's Keramid 601 polyimide resin are fabricated in a conventional laminating press, and processed in a manner similar to that used for epoxies but with an extended cure cycle or post-cure. The room-temperature mechanical and electrical properties are similar to epoxy laminates, as shown in Table 9.4. At elevated temperatures, the polyimides exhibit exceptional stability. In particular, the thermal coefficient of expansion in the Z axis does not change significantly up to approximately 240°C, as shown in Fig. 9.11. Exhaustive tests have shown[8] that polyimide-based multilayer boards can withstand repetitive thermal cycling at elevated temperatures (>150°C) without cracking of plated through holes. Similar excellent results were also obtained in solder shock tests (10 s at 288°C in molten solder). The thermal stability of these materials is retained at temperatures of approximately 200°C for continuous exposure in air, which has qualified them for military applications.

Polyimides are also being used in conjunction with silica fabric (which has a coefficient of expansion approximately one tenth that of E-glass) to produce laminates with expansion coefficients in the XY plane which are as low as 6–7 ppm, i.e. close to that of a ceramic chip carrier. This composite satisfies the requirement for a substrate material for *direct* attachment of the larger ($\frac{1}{2}$ to 1 in dimension) leadless chip carriers which are used for mounting LSI and VLSI chips.[9]

Although polyimides are being employed increasingly in printed wiring applications at the leading edge of the technology, they are four to five times more costly than epoxy resins and are more difficult to process. Polyimides account for only 1% of the total laminate resin consumption in the USA. Despite high costs, their use is expected to grow rapidly.

Table 9.4

Properties of Printed Wiring Laminates Based on Different Polymers

Property	Test	FR4 epoxy 30–35% resin	Keramid 601 polyimide 22–25% resin	FR2 phenolic 40–50% resin	Teflon glass-fabric
Flexural strength (psi×10⁻³) 25°C	A	80	71	20	15
299°C		30	57		
Flexural modulus (psi×10⁻⁶)	A	2·5	4·0	1·2	
Interlaminar bond strength	A	2·5	2·1	1·0	0·8
Dielectric constant (1 MHz, 25°C)	D24/23	4·3	4·6	4·0	2·35
Dissipation factor (1 MHz, 25°C)	D24/23	0·019	0·018	0·03	0·0005
Volume resistivity (×10⁻¹⁵ Ω cm)	C96/90/35	0·2	0·15	0·01	0·2
Electric strength (kV/mm)	A	22	20	25	28
Deflection temperature (°C)(264 psi)	E1/150	120	240	60	250
Thermal expansion (XY plane, ppm/°C)	E1/150	12	11	18	20
Water absorption (%) (24 h)	D24/23	0·1	0·3	0·6	0·02
Maximum continuous temperature (°C)		150	250	105	250

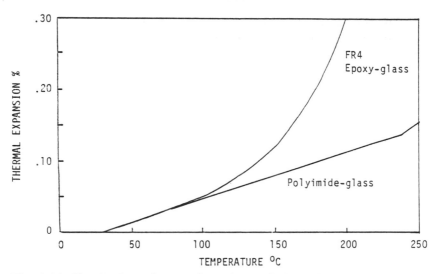

Fig. 9.11. Z axis thermal expansion of a typical polyimide-glass laminate compared with FR4 epoxy.

9.6.3. Phenolic Resins

Phenolics were the first thermosets to be used for the fabrication of electrically insulating substrates because of their ease of processing and excellent chemical and heat resistances.

Phenolic or phenol–formaldehyde resins describe a variety of products made by the condensation of phenols with formaldehyde. The first commercial processes for the manufacture of phenolic resins were developed by Baekeland and Thuslow in the early 1900s and the products were sold under the trade names of bakelite and novolac.

Two basic types of reaction occur between phenol and formaldehyde depending upon the pH of the reaction medium. In the presence of an acid catalyst, formaldehyde forms a carbonium ion, namely

$$CH_2 = O + H^+ \rightleftharpoons {}^+CH_2 - OH$$

which can add (electrophilically) to either of the two ortho or the para positions of a phenol

Since the linking of the phenolic groups may also occur at either the ortho or the para positions, branching occurs as the polymerisation proceeds

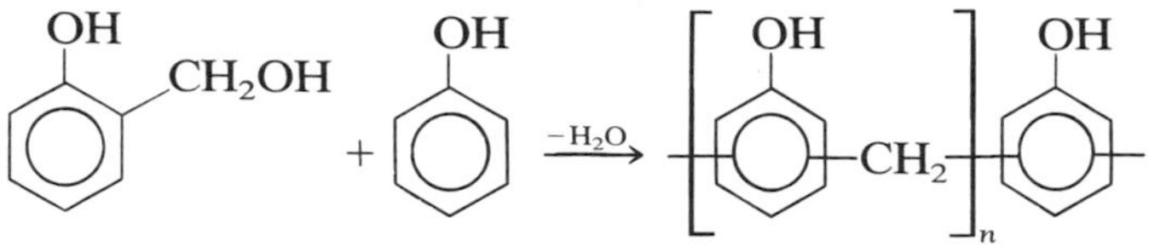

$$
\text{(phenol-CH}_2\text{OH)} + \text{(phenol)} \xrightarrow{-H_2O} \left[\text{(phenol)}-CH_2 \right]_n \text{(phenol)}
$$

The resulting *novolac* resins (prepolymers) generally have a value of n in the range 6–12.

If the reaction is base catalysed and excess formaldehyde is present, the condensation primarily occurs through formation of ether linkages

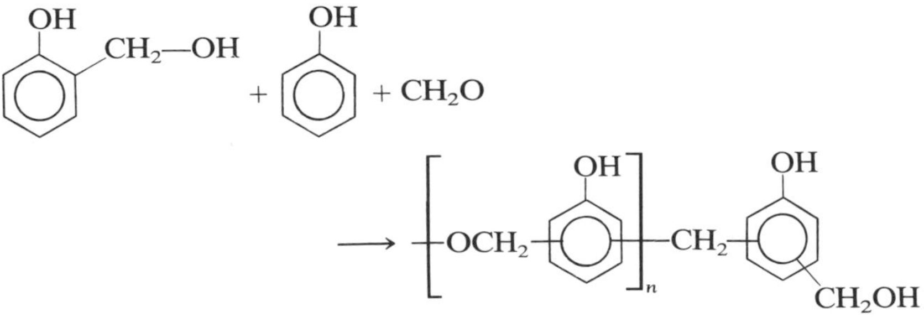

$$
\text{(phenol-CH}_2\text{—OH)} + \text{(phenol)} + CH_2O
$$

$$
\longrightarrow \left[OCH_2-\text{(phenol)} \right]_n -CH_2-\text{(phenol-CH}_2\text{OH)}
$$

The resulting *resole* resins can undergo further (thermal) reaction with elimination of water to produce a highly crosslinked system. By contrast, curing of the novolac resins requires the addition of a curing agent capable of forming additional methylene groups. Curing agents most commonly used are resole resins or hexamethylene tetramine. In the case of the latter, ammonia rather than water is liberated in the condensation reaction. Since novolacs require a second stage to fully cure, they are frequently referred to as two-step phenolics, whereas the resoles are referred to as one-step phenolics.

Phenolic resins used in lamination are primarily the one-step resoles, although novolacs may be added to improve resistance to water. Further improvements in water resistance and a reduction in brittleness are achieved by using para-cresol in place of phenol. This inhibits branching since the para position is blocked by the methyl group.

Phenolics are the lowest cost resins used in printed wiring substrate fabrication, but in the USA; account for only 3% of the volume due to their replacement by higher performance epoxies. In Japan, phenolic resins represent over 50% of consumption in printed wiring due to the wide acceptance of XXXP and FR2 grade laminates for consumer electronics. The properties of typical materials are shown in Table 9.4.

The major disadvantage of phenolics is low insulation resistance. Phenolics generally exhibit greater water absorption and sensitivity of the electrical properties to humid environments than epoxy resins, although only the so-called low-loss grades are used for printed wiring substrates. For these reasons, the use of phenolics in laminate fabrication has been restricted to the lower cost paper-based reinforcements.

9.6.4. Polytetrafluoroethylene

The use of polytetrafluoroethylene or PTFE for printed wiring substrate fabrication is a relatively well-established practice brought about by the need for improved high-frequency materials. The critical properties in this field of application are a low dissipation factor, a uniform dielectric constant which exhibits little variation with frequency over a wide bandwidth up to 15 GHz, combined with environmental and temperature stabilities. These properties can be achieved in a number of teflon/glass or teflon/glass/ceramic composites based on both fabrics and random glass. The dissipation factor at 1 MHz is more than 100 times less than a typical epoxy FR4, as shown in Table 9.4.

Polytetrafluoroethylene is the polymer resulting from the free radical polymerisation of tetrafluoroethylene $CF_2{=}CF_2$. The reaction is usually carried out in an aqueous emulsion which produces sub-micron particles of the polymer, or in a suspension which gives rise to particles of 100 μm or greater. The polymer is processed from various resin grades, which differ in fineness, by means of preforming followed by sintering as in powder metallurgy.

The major disadvantage of teflon is its high cost (approximately 20 to 30 times greater than FR4 laminate); however, no material is currently available which can compete with it for exacting microwave applications.[10] Some of the high temperature thermoplastics described in the next section can be used in its place for the fabrication of high-frequency substrates. Other applications of fluorinated polymers are discussed in Chapter 10.

9.7. THERMOPLASTICS FOR MOULDED AND EXTRUDED SUBSTRATES

Electronic grade thermoplastics, such as polyamides (nylons), polycarbonates, polystyrenes and polyphenylene oxides, are widely used in the electronics industry for the fabrication of housings, connectors, stand-offs and many other insulating and dielectric parts. Most of these polymers are unsuitable as printed wiring substrate materials because of poor chemical resistance and thermal stability. There are, however, a number of high-performance (engineering grade) thermoplastics, most of which have been developed during the past 10 to 15 years, that possess the required physical and chemical properties to compete with laminates in the printed wiring substrate forum.

In particular, these materials can be extruded and injection moulded offering the potential for low cost volume production. In the case of moulded boards, the advantages include the ability to mould design features such as holes, chip carrier cavities and standoffs. The three most important classes of high temperature thermoplastics are polysulphones, polyphenylene sulphides, and the very recently intro-duced polyetherimide (Ultem).

9.7.1. Polysulphones
Polysulphones are characterised by the presence of the following structural element in the repeat unit

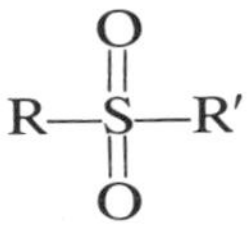

$$R-\overset{\overset{\displaystyle O}{\|}}{\underset{\underset{\displaystyle O}{\|}}{S}}-R'$$

In all of the high temperature polysulphones, R and R' are aromatic groupings.

Udel® polysulphone was the first polysulphone to be introduced commercially by Union Carbide in 1965. It is prepared by the reaction of the disodium salt of bisphenol A with dichloro-diphenylsulphone

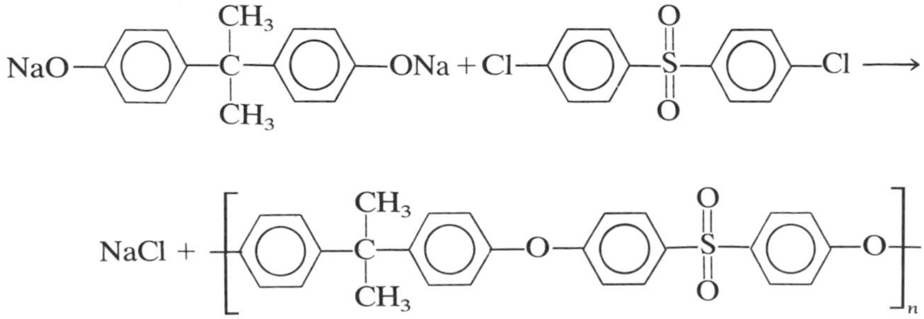

The reaction is carried out at 150°C in dimethylsulphoxide and leads to a value of $n > 100$.

Victrex® polyethersulphone was introduced in 1972 by ICI. It is prepared by the self-condensation of 4,4'-dichlorodiphenylsulphone in a strong base using a high boiling point solvent.

Union Carbide introduced Radel® polyphenylsulphone in 1976 but details of its manufacture are not currently disclosed.

In all these materials, the presence of the diarylsulphone grouping enhances the stabilisation of the S–C bonds through conjugation with the aromatic groups, leading to glass transition temperatures above 200°C. These bonds are also hydrolytically stable conferring strong resistance to attack by acids and alkalis. Furthermore, since the S atom is in its highest oxidation state, it is not susceptible to oxidative attack. The ether linkage imparts flexibility to the polymer chain and hence toughness to the material. The basic properties of the polysulphones are shown in Table 9.5 in comparison with other high temperature plastics. In general, all the materials are transparent and exhibit high rigidity, low creep and excellent thermal stability and flame resistance in comparison with other thermoplastic materials. For printed wiring substrate applications, the resins are usually reinforced with 20–30% glass or glass/mineral combinations to increase the rigidity (flexural modulus) and dimensional stability.

These plastics exhibit low dielectric losses and stable dielectric constants over a broad range of temperatures and frequencies. The chemical resistance of the materials is their weakest point. Although resistant to most common solvents, acids and alkalis, they exhibit stress cracking in the presence of organic ketones, esters and chlorinated hydrocarbons. Polysulphones are used in the electronics field for connectors, chip carriers and capacitor dielectrics, and they are the first of the high temperature thermoplastics to be used for printed wiring board fabrication.

9.7.2. Polyphenylene Sulphide

Polyphenylene sulphide or PPS contains the p-phenylene sulphide grouping as the repeat unit of the polymer chain. It is prepared by the reaction of para-dichlorobenzene with sodium sulphide in a polar solvent

$$\text{Cl}-\!\!\!\bigcirc\!\!\!-\text{Cl} + \text{Na}_2\text{S} \longrightarrow \left(\!\!\bigcirc\!\!-\text{S}\!\right)_n + 2\text{NaCl}$$

Table 9.5
Properties of High Temperature Thermoplastics

Property	Polysulphone Udel	Polyether sulphone Victrex	Polyphenylene sulphide Ryton	Polyetherimide Ultem	Ultem +20% glass reinforced
Flexural strength ($\times 10^{-3}$ psi)	15·4	18·6	25	21	30
Flexural modulus ($\times 10^{-6}$ psi)	0·39	0·37	1·7	0·48	0·9
Tensile strength ($\times 10^{-3}$ psi)	10·2	12·2	16·2	15·2	20
Dielectric constant (1 MHz, 25°C)	3·03	3·45	3·8	3·1	3·5
Dissipation factor (1 MHz, 25°C)	0	0·008	0·0014	0·006	0·0015[a]
Volume resistivity ($\times 10^{-15}$ Ω cm)	50	100	45	6·7	0·7
Electric strength (kV/mm)	17	16	17·7	28	26·5
Deflection temperature (°C) (264 psi)	174	202	243	200	209
Thermal expansion coefficient (ppm/°C)	56	55	40	56	25
Water absorption (%) (24 h)	0·3	0·4	0·05	0·25	0·26
Maximum continuous temperature (°C)	160	170–200	170	170	170

[a] 1 kHz

The product which is essentially a prepolymer, is a linear, relatively low molecular weight material with $n = 170$, a glass transition temperature of 85°C and a melting point of 285°C. Further curing in air is required either by heating the solid at temperatures below the melting point or by heating the melt. The curing process involves partial oxidation which results in chain extension and some crosslinking. Cured compositions exhibit heat deflection temperatures in excess of 140°C. In this respect, PPS is more like a thermoset than a true thermoplastic material; however, it is generally processed (injection moulded) in the thermoplastic form and the final cure does not involve pronounced changes in properties such as high exotherms, shrinkage and elimination of volatiles as occur with many conventional thermosets.

The basic properties of cured PPS moulding compositions are shown in Table 9.5. The most distinguishing feature of the materials is their outstanding resistance to almost all organic solvents, even at elevated temperatures, and their high thermo-oxidative stability. The materials are also inherently flame-resistant. On account of its excellent electrical insulation properties and general stability coupled with the ease with which melt viscosity can be varied, Ryton PPS is widely used for the injection moulding of electrical connectors, and has begun to find its way into printed wiring board substrates since it can compete with other high temperature thermoplastics on the basis of materials fabrication costs and chemical stability. The use of polyphenylene sulphide as a component encapsulant is discussed in Chapter 10.

9.7.3. Polyetherimide (Ultem®)

Ultem was introduced by General Electric in 1982 and is described as a thermoplastic member of the polyimide family. Details of its structure and method of production are not yet published. The properties of the glass-reinforced moulding compound are shown in Table 9.5. It is evident that this material has electrical and mechanical properties very similar to those of the other high temperature thermoplastics. Chemical resistance is inferior to PPS but similar to the polysulphones. The thermal expansion of a glass-filled composition is essentially isotropic and, as with the other high temperature thermoplastics, exhibits significantly less temperature dependence in the Z axis compared with epoxy FR4. The material can be readily processed by extrusion or injection moulding but must be thoroughly dried before use, indicating a potential sensitivity to water. It is currently under evaluation by a number of companies for printed wiring substrate applications.

9.7.4. Fabrication and Uses

High temperature thermoplastics can be extruded to produce both copper-clad and unclad boards. The latter are functionalised with an adhesive coating for additive or semiadditive processing (see Section 9.4.4).

Moulded boards must also be processed by additive or semiadditive techniques and a number of proprietary methods have been developed which are based on modifications of laminate additive technology. Moulded board fabrication requires high moulding pressures (>5 tons/in^2 for Ultem) which restrict the maximum size of substrate to about $10'' \times 10''$ using a 500 ton injection moulding machine.[11] Tooling costs will limit the number of units per run to 100 000.

Although the use of high temperature thermoplastics currently accounts for less than 1% of the substrate market, this small base is expected to grow significantly. One reason for this predicted growth, aside from the potential economic advantages of moulded or extruded boards, stems from the low dielectric constant and loss exhibited by the high temperature thermoplastics compared with high glass content laminates.[12] This property, coupled with good thermomechanical stability, has generated considerable interest in their use for high-frequency applications as an alternative to more expensive materials such as teflon. Applications will involve high volumes of identical substrates, in which design flexibility is important. This could include telecommunications, automotive and consumer electronics. A number of commercial applications are described in a recent review.[13]

9.8. POLYMERS FOR FLEXIBLE PRINTED WIRING

Flexible printed wiring is used primarily in electronic packaging applications where low weight or volume is a prerequisite, or excessive vibration is likely to be encountered. The main advantage of flexible circuitry is its ability to be shaped into more than one plane or conform to an irregular shaped package.

The development of flexible printed wiring was made possible through the availability of plastic films that possess high electrical insulation and good chemical resistance, in particular polyethylene terephthalate and polyimide films which account for over 95% of the flexible film consumption for printed wiring applications. The properties of the most important of these polymers are discussed in the following sections.

9.8.1. Polyethylene terephthalate (PET)

Polyethylene terephthalate is the linear polycondensation product of ethylene glycol and terephthalic acid

$$H\left[OCH_2CH_2O-\overset{\overset{O}{\|}}{C}-\overset{}{\bigcirc}-\overset{\overset{O}{\|}}{C}\right]_n OCH_2CH_2OH$$

It was introduced commercially in fibre form in the 1950s. The polymer is processed into film by melt extrusion followed by hot rolling and stretching which results in molecular orientation accompanied by partial crystallisation. The deformation-induced orientation imparts high strength and toughness to the film which is not a characteristic of the purely amorphous phase. PET film is available in a number of grades for a variety of applications including magnetic tapes, photographic substrates, packaging and electrical insulation. Mylar® Type A from DuPont or equivalent grades from ICI and Hoechst are normally used for flexible printed wiring and the typical properties are shown in Table 9.6. The outstanding features of Mylar are its high dielectric strength, low dissipation factor and excellent resistance to chemicals and water. It also possesses a high tensile strength and does not embrittle with ageing. The main limitation of Mylar is its inability to withstand temperatures above 150°C and hence it has very poor solderability. However, because of its low cost it is used extensively in automotive dashboards where crimped terminations can be used in place of soldered joints because of the small number of components to be interconnected and the low interconnection density. It is also used purely as a flexible harness in automotive and related applications.

9.8.2. Polyimide

Only one type of electrical grade polyimide is available in film form and this is manufactured by DuPont under the trade name Kapton®. Kapton polyimide is formed by the polycondensation of pyromellitic dianhydride and 4,4'-diamino diphenylether. The reaction proceeds via the intermediate formation of the polyamic acid as described in Section 9.6.2.

The polymer is melt-processed into the film form, the properties of which are shown in Table 9.6. Kapton film possesses excellent dielectric and insulation properties which remain constant over a wide range

Table 9.6
Properties of Plastic Flexible Film Materials

Property	Mylar polyester type A 1 mil	Kapton polyimide type H 2 mil	Dacron-epoxy R/2400 4 mil	Nomex nylon type 410 5 mil
Tensile strength ($psi \times 10^{-3}$)	20	25	5·5	11
Tear strength ($lb/in \times 10^{-3}$)	2	0·9	5·3	6·5
Folding endurance (MIT, $cycles \times 10^{-3}$)	14	10	50	5
Dielectric constant (1 MHz, 25°C)	3·0	3·5	$3·2^a$	$2·3^a$
Dissipation factor (1 MHz, 25°C)	0·016	0·0025	$0·015^a$	$0·01^a$
Volume resistivity ($\Omega\,cm \times 10^{-5}$)	10^3	10	10^{-3}	20
Electric strength (kV/mm)	280	280	36	30
Shrinkage (%/min at 150°C)	2–3	0·1		$0·9^b$
Thermal expansion coefficient (ppm/°C)	27	20	20	
Water absorption (%) (24 h)	0·01	2–3	1–1·5	5
Maximum continuous temperature (°C)	100	250	100–150	220–250

^{a} 1 kHz.
^{b} 300°C.

of temperature and frequency. The outstanding feature of Kapton film, in keeping with other polyimides, is its high thermal stability and its ability to retain good physical properties over a wide temperature range (-250 to $+250°C$). The main disadvantage of polyimide film is its relatively high water absorption which results in an increase in the dielectric constant and dissipation factor and a reduction in the resistivity and dielectric strength on exposure to high-humidity environments. Moisture must be removed from Kapton films prior to lamination with copper foil to avoid blistering and delamination.

Since polyimide film possesses excellent dimensional stability, high tensile strength and high temperature resistance, its use in more sophisticated, flexible circuit applications is growing rapidly. Polyimide film is also used for the fabrication of a hybrid type of wiring board referred to as 'flex-rigid'[14] in which conventional rigid glass reinforced laminate wiring boards are combined in a multilayer package with integral printed flexible cables that are electrically connected by means of plated through holes. This type of circuit board is used mainly for military and high reliability sytems.

9.8.3. Plastic Composites

Two composite materials which are also used for low-cost flexible wiring applications are Dacron-epoxy® and Nomex® by DuPont. Dacron-epoxy consists of a non-woven polyester fibre mat embedded in an epoxy resin. The properties of the material are shown in Table 9.6. It possesses good chemical and moisture resistances, excellent dimensional stability and has the highest folding endurance of all of the currently used flexible film materials. It is also solderable and flame resistant and has a cost similar to Kapton. Nomex is a low-priced nylon paper composite based on an aromatic polyamide which possesses higher temperature resistance than conventional polyamides. Nomex paper is composed of two different forms of the same polymer, consisting of short fibres which provide high tensile strength and sub-micron fibres which act as a binder. The properties of the calendered type Nomex 410 are shown in Table 9.6 in comparison with other flexible film materials.

Both Dacron-epoxy and Nomex are solderable and flame resistant. Nomex, however, possesses much higher water absorption and is therefore not generally used for applications involving humid environments.

9.8.4. Fabrication and Uses

Flexible film substrates are available as single- and double-sided copper-clad laminates. Copper foil is generally laminated to the base material by means of epoxy- or acrylic-based adhesives.

Flexible printed wiring can be fabricated by means of either the additive or the subtractive process as used for rigid substrates. An alternative method which is growing in use because of its low cost is the screen printing of low temperature curable conductive inks directly onto the base material (usually Mylar). This additive technique is being used for the fabrication of single-sided keyboards. Other low-cost consumer electronic applications for flexible wiring include cameras, microwave ovens, video games, watches and calculators. More sophisticated instrumental and computer applications are described in a recent review.[15]

9.9. SUMMARY AND CONCLUSIONS

The most widely used plastic materials in the printed wiring industry are epoxy resins based on bisphenol A. Epoxy-glass fabric laminates account for approximately 80% of the consumption of printed wiring substrates in the USA. In particular, laminates with specifications equal to or better than NEMA grade G10 or FR4, which combine very good dimensional stability with excellent electrical properties, are suitable for the fabrication of high density wiring that is required for computers and telecommunications equipment.

In the consumer electronics field where interconnection density is generally low, cheaper substrates such as paper-based phenolic resins, epoxy-glass mat composites and polyester or nylon based flexible films are employed.

Polyimides which are available as both laminating resins and flexible films are used in military and aerospace applications where high temperature performance is critical.

High temperature thermoplastics such as polysulphones and polyetherimides are newcomers to the printed wiring substrate arena and offer the potential for the production of relatively low-cost moulded substrates with excellent electrical characteristics and thermal stability. The low dissipation factors characteristic of these materials have generated considerable interest in their potential for use in high-frequency circuits, as a substitute for the much more expensive teflon-based substrates. Much of the demand for new or improved substrate materials for printed wiring is generated by the ever-increasing density of circuitry associated with LSI and VLSI microchips. As the chips become larger and more functional the number and density of interconnections increase. This requires the use of finer traces, smaller pads and plated through holes, in conjunction with an increasing employment of multilayer wiring. Consequently, substrates with high dimensional stability are required to facilitate the precise registration of circuit layers and elimination of failures due to the cracking of plated through holes. This latter results from excessive expansion and contraction of the substrate that occurs during thermal cycling in such processing steps as lamination and soldering. Furthermore, the higher current densities associated with LSI and VLSI chips give rise to a greater level of heat dissipation and hence thermal stress which adds to the problems of substrate thermomechanical stability.

Today, most of these problems are being resolved by the use of rigid substrates based on high glass transition temperature resins such as polyimides and modified epoxides. The reduced thermal expansion of these materials at the normal processing or operating temperatures provides for significant increases in dimensional stability compared with G10 or FR4 laminates.

In addition, composites are being developed in which the thermal coefficients of expansion are matched more closely to those of the chip substrates and copper foils to reduce thermal stress. Many of these developments at the so-called leading edge of the technology are heralding the future for printed wiring substrates. What is certain is that no single plastic material or composite will provide the necessary cost/performance capability for all the increasingly diverse needs of the electronics industry.

ACKNOWLEDGEMENTS

Sources of Data Used in Tables and Figures
1. Technical bulletins and product brochures published by:

Dow Corning	Epoxy resins
Ciba Geigy	
Rhone Poulenc	Keramid

DuPont	Kapton
	Mylar
	Nomex
	Dacron-epoxy
General Electric	Ultem
Union Carbide	Udel
Phillips Chemical	Ryton
Rogers Corporation	PTFE Composite
Dynacircuits	NEMA Laminates
U.O.P.	NEMA Laminates

2. *Plastics for Electronics, Desk Top Data Bank*, International Plastics Selector Inc., 1979.
3. Lee, H., and Neville, K., *Handbook of Epoxy Resins*, McGraw Hill, New York, 1967.

FURTHER READING

Coombs, C. F., *Printed Circuits Handbook*, McGraw Hill, New York, 1979.

Harper, C. A., *Handbook of Materials and Processes for Electronics*, McGraw Hill, New York, 1984 (1970 reissue).

Swartz, S. S. and Goodman, S. H., *Plastics Materials and Processes*, Van Nostrand, New York, 1982.

REFERENCES

1. Siegmund, J., Private Communication, Siegmund Inc., 1984.
2. Bubb, J. R. *et al.*, *IBM, J. Res. & Develop.*, **26**(3) (1982) 306–16.
3. Business Communications Inc., Stanford, CT, USA; Gnostic Concepts Inc., Menlo Park, CA, USA; BPA (Technology and Management Ltd), Dorking, England.
4. IPC (Institute for Interconnecting and Packaging Electronic Circuits), Evanston, IL, USA.
5. Knox, C. E., Fibreglass reinforcement. In: *Handbook of Composites*, (Ed. G. Lubin), Van Nostrand, New York, 1982. Kesell, S. L. M., Glass fibers for the printed circuit industry, *Proc. National Electronic Packaging and Production Conference*, Anaheim, 1981, Cahners Exposition Group.
6. Haug, T., New laminating resins with improved thermal, mechanical and chemical properties, Paper WC111-53, *Printed Circuit World Convention III*, Washington DC, May 1984; Azuma, K., Epoxy resin laminates based

on oxazolidine curing systems, ibid, Paper WC111-19; Urscheler, R., New epoxy laminating system with high heat distortion temperature, ibid, Paper WC111-19.

7. Serafini, T. T., High temperature resins. In: *Handbook of Composites*, (Ed. G. Lubin), Van Nostrand, New York, 1982.

8. Zeroth, J. C., Polyimide materials gaining new ground, *Electronic Packaging and Prod.* (July 1981) 113–23.

9. Gates, L. E., Jr. and Reimann, W. G., Quartz fiber in PCB's improves temperature stability, *Electronic Packaging and Prod.* (May 1983) 68–73.

10. Miller, T. L., Teflon substrates for microwave circuitry, *Circuits Manuf.*, (May 1984) 73–5.

11. Rush, O. W., Plastic injection molding tackles substrates, *Electronic Packaging and Prod.* (Oct. 1983) 92–5.

12. Engelmaeir, W. and Frisch, D. C., Injection molding shapes new dimensions for boards, *Electronics* (Dec. 1982).

13. Galli, E., Molded thermoplastic P.W.B.'s, *Plastics Design Forum* (May/June 1984) 21–30.

14. Zawicki, E. A., Flex-rigid packaging—an overview, *Proceedings of 4th International Printed Circuit Conference*, NY, 1981.

15. Lyman, J., Flexible circuits bend to designer's will. In: *Microelectronics Interconnection and Packaging* (Ed. J. Lyman), McGraw Hill, New York, 1982, 116–24.

Appendix A1
Summary Properties of Polymers Used in Electronics

	Volume resistivity ($\Omega\ cm \times 10^{-15}$)	Dielectric constant (1 kHz)	Dissipation factor ($1\ kHz \times 10^3$)	Heat distortion temperature (°C)	Maximum continuous service temperature (°C)	Water absorbance (24 h) (%)	Chemical and solvent resistances	Price range per pound[a]
Thermoplastics								
Acetal	0·1	3·7-3·8	4-5	125	85	0·25	Fair	M
ABS	1-100	2·5-4·5	4-7	80-120	70-115	0·2-0·5	Poor	L
Acrylic	0·1-10	3·3-3·5	40-60	70-100	60-95	0·3	Poor	L-M
Nylon 6	0·01-0·1	3·6-4·0	14-20	65-105	120	1·3	Good	M
Polyamide-imide	100	3·5	1	275	290	0·28	Good	H
Polyarylether	15	2·9	4	150	120	0·25	Excellent	M
Polycarbonate	10	2·9-3·0	20	130-140	120	0·15	Fair	MH-H
Polyethylene HD	10-10³	2·3	0·2-0·5	45-50	80-105	<0·01	Good	L
PET	>10³	3·0-3·1	5	230	150	0·6	Good	L
PPO	100	2·6	0·4	100-130	80-105	0·06	Good	MH
PPS	50	3·1	0·4	245	260	0·02-0·03	Good	H
Polypropylene	0·01-100	2·2-2·6	0·5-1·8	50-60	105-125	<0·01-0·03	Good	L
Polystyrene	>100	2·4-2·7	0·1-0·3	80-105	60-115	0·03-0·05	Poor	L
Polysulphone	50	3·1	10	175	150-175	0·2-0·3	Poor	MH
PTFE	>10³	2·1	0·2	55	285	0·01	Excellent	H
PVC	10⁻³	4-8	80-150	60-80	75	0·15-0·75	Fair	L
SAN	1	2·6-3·3	7-10	90-105	95-105	0·23-0·28	Fair	L
Thermosets								
Diallyl Phthalate	>100	3·3	8	240	150-175	0·15	Good	MH
Epoxy	0·1-10	3·3-3·7	2-20	45-285	120-285	0·02-0·03	Excellent	M
Melamine	0·01	6·6	0·02-0·03	135	100-200	0·6	Good	L
Phenolic	10⁻⁴-0·01	4-5·5	10-50	150-200	150-250	0·2-1	Good	L
Polyimide	10-100	3·4	5	240-360	260	0·32	Excellent	H
Polyurethane	10⁻¹-50	3·4-3·5	5	90-95	95	0·1-0·2	Poor	L
Silicone	0·1-1	2·6-2·7[b]	1-2[b]	—	260	<0·1	Good	MH

[a] Price range: L= below 50¢; M= 50¢-$1; MH= $1-$2; H= above $3.
[b] 1 MHz.

R. Hurditch

Appendix A2

Summary Properties of Some Materials Used in the Fabrication of Electronic Circuits

	Dielectric constant (1 MHz)	Dissipation factor (1 MHz)	Thermal expansion coefficient (ppm/°C)	Thermal conductivity (cal/°C/cm/s $\times 10^3$)
Borosilicate–E-glass	6·33	0·001	5·04	3·0
Fused silica	3·7	0·000 1	0·54	2·0
Aluminium oxide 99·5% (sintered)	9·4	0·000 1	6·3	75
Beryllium oxide 99·5% (sintered)	6·8	0·000 3	6·4	670
Epoxy resin	3·3–4·0	0·03–0·05	48–65	0·75
FR4 Laminate	4·3	0·019	10–12 (XY plane)	1–1·5
Copper	—	—	16·7	770
Aluminium	—	—	7·2	460
Silicon	—	—	2·3	350

Chapter 10

OTHER MATERIALS AND THEIR APPLICATIONS IN ELECTRONICS

MARTIN T. GOOSEY

Dynachem Corporation, Tustin, California, USA

10.1. INTRODUCTION

In addition to the materials and applications covered in detail in the preceding chapters, there are a large number of other polymeric materials and applications for polymers in the electronics industry. Many of these applications involve relatively new and novel materials or materials with a unique combination of properties that make them particularly useful in a specific area. In some cases new plastic materials are making feasible previously impossible processes or processes which were only possible using more complex and less efficient materials and equipment. The evolution of these new materials also offers the promise for further improvements in numerous products giving enhanced properties when compared to the same products made with older generation materials. This final chapter attempts to briefly cover some of these new materials and applications. Also covered are some other well-known and some older polymers which, although being widely used in other industries, have not been used sufficiently in the electronics industry to warrant a whole chapter in this work. The polymers and applications covered here nevertheless represent an important part of the whole story of the use of plastics in electronics and where possible references are cited to guide the reader towards further information.

10.2. LOW STRESS AND TOUGHENED EPOXIES

With the increased levels of integration now being used in microelectronic devices, the width and spacing of the aluminium conductors deposited on the chip surface are becoming smaller. Coupled with this, the increased size of both chips and packages has meant that some assembled devices have become susceptible to internal stresses. These stresses can deform the aluminium conductors on the device surface causing problems such as short circuiting of adjacent conductors and rendering the metal susceptible to stress corrosion. Passivation layers can also be damaged by these stresses, allowing moisture to permeate to the underlying metal, again increasing the likelihood of early corrosion failure. In the worst cases the die or even the package itself can crack.

The principal cause of internal stresses in plastic encapsulated devices is the large difference in physical properties between the moulding compound and the encapsulated leadframe and device. Various parameters affect the internal stress characteristics of a moulding compound and, in particular, the thermal expansion coefficient, glass transition temperature, viscoelasticity, Young's modulus and degree of stress relief attainable are thought to be important. One of the major factors influencing both the Young's modulus and the thermal shrinkage properties of a moulding compound is the filler loading. In order to reduce the internal stress a material will impart upon a device the Young's modulus and thermal expansion coefficients must both have low values. Unfortunately it is not possible to reduce the thermal

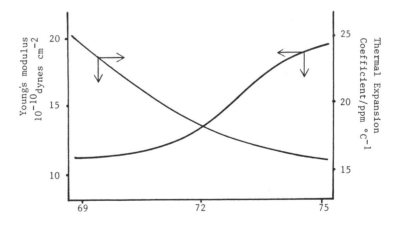

Filler/% weight

Fig. 10.1. Variation of Young's modulus and thermal expansion coefficient with filler content for an epoxide moulding compound.[1]

expansion and the Young's modulus at the same time, since a reduction in filler content, whilst reducing the Young's modulus, increases thermal expansion coefficient, as shown in Fig. 10.1.

There is thus an optimum filler loading at which minimum internal stress performance is achieved and by utilising this fact it is possible to significantly improve the stress performance of epoxide moulding compounds. The properties of two commercially available low stress epoxide moulding compounds are compared to a standard material in Table 10.1.

In order to improve the internal stress performance even further or to provide toughness for potting and adhesive applications it is advantageous to introduce reactive additives to the basic epoxide compound formulation.[2] One of the most popular additives that has found considerable use in this type of application is a rubber-like liquid copolymer known as carboxyl terminated butadiene/acrylonitrile (CTBN). The generalised molecular structure of this material is:

$$HOOC\text{---}\left[(CH_2\text{---}CH\text{=}CH\text{---}CH_2)_x\text{---}(CH_2\text{---}\underset{\underset{\displaystyle C\text{=}N}{|}}{CH})_y\right]_z\text{---}COOH$$

where x, y and z would typically be in the region of 5, 1 and 10, respectively. The carboxylic acid groups at each end of the molecule (carboxyl termination) provide sites for reaction with the epoxide resin and so the CTBN can be chemically bonded into the curing polymer matrix. The extent of this reaction and the final cured properties are very dependent upon resin type, stoichiometry and the catalyst used but the resultant product is considered to be a solution of minute rubber particles usually less than 1 μm in diameter in the epoxide and

Table 10.1
Properties of Low Stress Epoxide Moulding Compounds[a]

	Thermal expansion coefficient (α_1, ppm/°C)	T_g (°C)	Internal stress (psi)
Reference	23	150	6800
Polyset® 420	20	160	5900
Polyset® 450	19	160	5400

[a] Data from the Polyset Division of the Dynachem Corporation.

these are easily identified by scanning electron microscopy. In addition to carboxyl terminated additives, hydroxyl, amine and vinyl terminations are also possible giving compounds identified as HTBN, ATBN and VTBN, respectively.

Care has to be taken when utilising these types of additive because some of the final cured properties may be degraded. Possible examples would be depression of glass transition temperatures and an increase in thermal expansion coefficients. However, by careful formulation, the advantages offered by CTBN-type additives can be made to far outweigh the disadvantages.

The electrical properties of a CTBN-containing epoxide may also be altered somewhat from those of the base material, but for most applications the effect is negligible. The dielectric constant of the cured compound will tend to be slightly higher and may show a greater dependence upon temperature than the unmodified compound, whilst the volume resistivity will tend to fall. The variation of electrical properties of a filled low density amine cured epoxide suitable for electrical applications is shown in Table 10.2 for both the standard and CTBN-modified materials.

For some electrical and electronic adhesive applications it may be advantageous to use CTBN modified epoxide adhesives in place of unmodified epoxides because of the enhanced tensile shear strengths achieved. The tensile shear strengths for a boron trifluoride amine complex cured DGEBA-type epoxide modified with various amounts of CTBN on solvent-cleaned cold-rolled steel are shown in Table 10.3.

A number of other toughening agents have also recently been investigated as alternatives to CTBN and its analogues. One reason for

Table 10.2

Variation of Electrical Properties for CTBN Modified and Standard Filled Low Density Amine Cured Epoxide[3]

Temperature (°C)	Dielectric strength (volts/mil)		Volume resistivity (ohm cm)	
	Standard	CTBN Mod.	Standard	CTBN Mod.
−54	589	384	$9 \cdot 8 \times 10^{14}$	$7 \cdot 2 \times 10^{15}$
24	296	347	$1 \cdot 1 \times 10^{14}$	$3 \cdot 9 \times 10^{14}$
93	302	328	$4 \cdot 6 \times 10^{13}$	$2 \cdot 3 \times 10^{13}$
204	124	115	$2 \cdot 3 \times 10^{10}$	—

Table 10.3
The Increase in Tensile Shear Strength with Addition of CTBN
for an Epoxide Adhesive[4]

Amount CTBN (%)	Tensile shear strength (psi)
0	1400
7.5	3900

this is that the unsaturated double bonds in the CTBN compounds may be susceptible to thermal or oxidative degradation, possibly leading to uncontrollable crosslinking and changes of properties. One group of compounds that may be alternatives to CTBN and which also offer better thermal stability are the elastomeric polysiloxane modifiers such as polydimethylsiloxane and polytrifluoropropylmethylsiloxane.

The rubber-like dispersion of particles throughout the polymer matrix that occurs when the above reactive liquid elastomers are added to an epoxide resin formulation provides toughening (increased resistance to thermal shock and fracture) without having a serious adverse effect upon the electrical properties. These additives are thus likely to become an increasingly popular tool for the compound formulator attempting to produce encapsulants and moulding compounds that allow devices to survive the stresses imparted during thermal and mechanical cycling or where low stress is required because of die size and packaging considerations.

10.3. LOW ALPHA PARTICLE EMITTING MOULDING COMPOUNDS

As dynamic memory devices have progressed from 8 and 16K to 64K and upwards, a phenomenon known as soft error generation has become an important problem in terms of reliable data storage. This is because of soft errors which are random, non-recurring, single bit errors occurring in memory devices such as dynamic random access memories (RAMs) and charge coupled devices. Dynamic memories store data as the presence or absence of electrons in the potential wells of the memory. Electrons can be generated by the interaction of alpha particles with the atomic structure of the chip. These electrons collect in the memory cells' potential wells thereby altering the memory state

Table 10.4

Actinide Content and Alpha Emission Rates for Various Components of an Epoxide Resin Moulding Compound[6]

Component	Uranium (ppb)	Thorium (ppb)	Alpha/h cm²
Fused silica	120–260	20–250	0·013–0·030
Epoxide resin	1	1	0·0001
Hardener resin	1	1	0·0001
Low alpha fused silica	1·3–1·5	0·7–2·0	0·00015–0·00022

and generating a so-called 'soft error'. As the levels of integration have increased, the size of the potential wells has decreased, effectively meaning that the difference in stored charge between memory states (the critical charge) is smaller. Fewer electrons are therefore required to alter a memory state and thus devices are becoming increasingly sensitive to alpha particle generated soft errors.

The alpha particles are radioactive decay products of actinide elements such as uranium and thorium, which are found in trace amounts in the materials used to fabricate both ceramic and plastic packages.[5] Of particular importance in the manufacture of moulding compounds for the plastic encapsulation of such memory devices are the inorganic components used in the formulations, because these are the principal sources of uranium and thorium. The epoxide resins and hardener resins are relatively free of these actinides and so in order to reduce alpha particle emissions in a compound it is necessary to improve the purity of the fillers. The alpha particle emissions and actinide concentrations for various components of a moulding compound are shown in Table 10.4. It is interesting to note that in order to achieve acceptable alpha emission rates for a material the permissible level of actinides must be as low as 1 ppb. By using raw materials mined from areas with inherently low levels of actinides, or by obtaining fillers via a synthetic route, it is possible to bring about a dramatic decrease in the alpha particle activity of moulding compounds. As a consequence, new low alpha emission moulding compounds, in combination with a better awareness of the design rules required to minimise the effects of alpha particles, will allow the continued use of plastic encapsulants in the packaging of highly integrated dynamic RAMs.

10.4. POWDER COATING

Since the development of the fluidised bed process, powder coating has become an important method of applying protective conformal coatings to electronic components. In the electrical and electronics industries thermosetting plastics and more specifically epoxides are the most widely used materials for making these films. For certain applications powder coating offers a number of advantages over other coating methods and these have contributed to the rapid growth in its use for encapsulating electronic components. Typical advantages are:

(1) a one-component material with a long shelf-life
(2) very good edge coverage and sag resistance
(3) excellent electrical, mechanical and chemical properties
(4) flexibility in processing
(5) low labour costs and moderate equipment costs
(6) no toxicity or flammable solvent problems.

The major use of powder coatings in the electronics industry is for the conformal coating of electronic components such as resistor networks, capacitors and hybrids.[7] Epoxide powder coatings have been formulated for application to components using automatic fluidised bed dipping equipment, often capable of handling in excess of 25 000 components an hour. Various other methods of application such as electrostatic spraying are also popular.[8] Powder technology overcomes several disadvantages of the older liquid dip systems in that the product is supplied as a single component material avoiding the need for the premixing of individual components. In addition, there are no pot life problems and cure times have been significantly reduced.

A suitable powder for electronic component coatings must usually have a fusing temperature well below the melting point of tin lead solder; 130°C is considered to be ideal. This gives ample latitude for a component preheating temperature of between 150 and 160°C which avoids the likelihood of solder reflow but also allows for some temperature drop between preheating and powder application. The materials used in the formulation of epoxide powder coatings for electronic component application are generally of the same purity as those used in transfer moulding epoxies ensuring device reliability. Also, because no mould release agents are required, powder coatings have excellent adhesion and give very good temperature cycling/thermal shock performance.

The particle size distribution of a powder coating is very important to the method of application. Fluidised bed operations require larger particle sizes than for the other common methods of application such as electrostatic spraying. For a fluidised bed, particles smaller than 175 μm with a balanced particle size distribution are required. In electrostatic coating, even smaller particles are used (<100 μm) because the charge imparted to a particle is proportional to its surface area (d^2). Since the mass of a particle is proportional to its volume (d^3), the smaller a particle is, the higher its charge-to-mass ratio will be. The properties of some typical powder coatings are shown in Table 10.5.

In the electrical industry the major applications of powder coatings are with electric motor manufacture and as wire insulators. Powder coatings are used as a replacement for paper slot liners, forming an integral coating over which the windings can be directly laid. They are also used for the encapsulation of end windings of motors for use in portable drills and other motor-containing domestically used equipment. Electrostatic powder coating offers an alternative to solvent-

Table 10.5
Typical Properties of Polyset® Powder Coatings[a]

Property	Compound		
	EPC 50	*EPC 20*	*EPC 46*
Hot plate gel time	20 s at 200°C	70 s at 150°C	40 s at 150°C
Plate flow	35 mm at 200°C	40 mm at 150°C	45 mm at 150°C
Application temperature (°C)	170–220	125–175	125–200
Shelf-life at 4°C (months)	12	12	12
Dielectric constant at 1 kHz	4·0	4·3	3·5
Dissipation factor at 1 kHz	0·008	0·0065	0·005
Suggested applications	Motor irons, meter coils	Tantalum capacitors R–C networks	Resistor networks Mica capacitors

[a] Data from the Polyset Division of the Dynachem Corporation.

Table 10.6

A Powder Coating Formulation Suitable for the Insulation of Wires[9]

Component	Parts by weight
Epoxide resin	800·0
Trimellitic anhydride	58·7
Flow control agent	13·6
Iron oxide pigment	16·0
Titanium dioxide pigment	8·0
Stannous octoate	1·6
Hydrophobic surface treated fumed silica	1·8

based enamels that have traditionally been used to insulate magnet wires. They have the advantage of being a solvent-free system that can be applied at the required thickness in a single operation. A typical powder coating formula suitable as a replacement for solvent-based enamels in the insulation coating of wires is shown in Table 10.6.

The use of properly applied and cured powder coatings can impart a number of important properties such as impact strength, abrasion resistance, moisture resistance, temperature cycling performance, electrical insulation characteristics and adhesion to a variety of substrate and device types, making them particularly useful in both the electrical and electronics industries.

10.5. POLYIMIDES

One of the main weaknesses of many polymeric materials is their poor elevated temperature performance, with softening and thermal degradation being the principal problems. In aerospace and military applications there is often a need for good high temperature stability and much research has been done on the development of heterocyclic and ladder-type polymers. One of the first types of high temperature polymer to become commercially important was the polyimides. They have been available since 1926, although not commercially until 1961, and have found considerable use in the electronics industry because of their good insulating and mechanical properties which are retained at temperatures in excess of 250°C.

Polyimides are a group of polymers containing highly aromatic structures, and it is this feature that is responsible for imparting their exceptional thermal stability. They are synthesised by the condensation reaction of a dianhydride and an aromatic amine, which yields a phthalimide grouping, the basic building block or repeat unit of the polymer chain

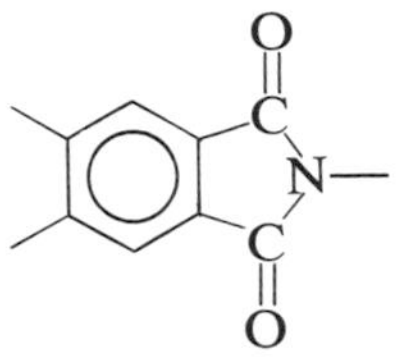

In order to achieve the high molecular weight, high thermal stability product a two-stage polymerisation reaction is performed. The first stage is the formation of an intermediate polyamic acid by the reaction of the dianhydride molecule with an amine group from the aromatic diamine. This intermediate polyamic acid is then made to undergo a ring closing condensation reaction with the elimination of water to give the fully cured, high molecular weight polyimide

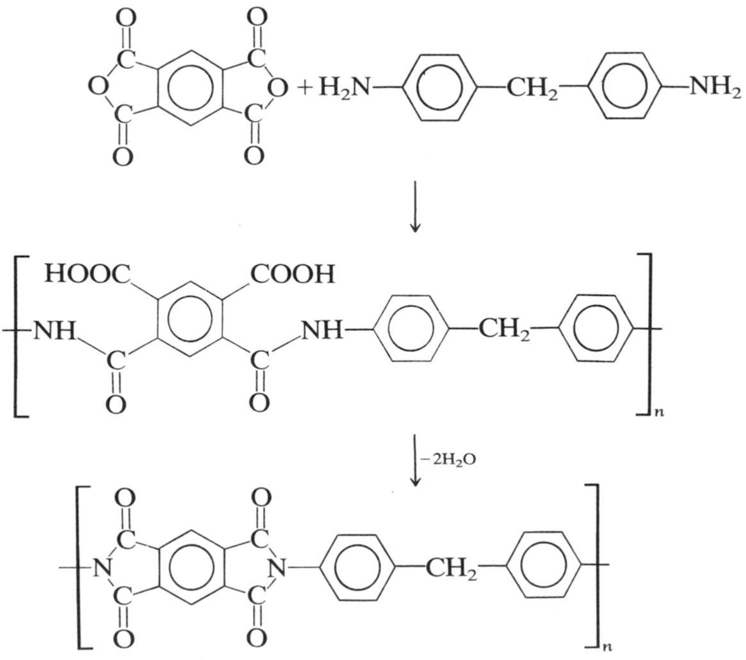

The ring closure condensation reaction is usually performed at temperatures of between 300 and 400°C and may take anything from several minutes to an hour depending upon the temperature.

These high temperatures have somewhat limited the use of polyimides to those applications where such temperatures can be tolerated. Recent work with compounds known as benzoguanamines and triazines has successfully lowered the final processing temperature to between 120 and 200°C. Another factor limiting the use of polyimides has been the insolubility of the fully imidised compounds. The solubility can be improved by making changes in the molecular structure such as the incorporation of aromatic polyether linkages into the polymer backbone or through the addition of pendant phenyl and alkyl groups. One other way of achieving enhanced solubility is to use the amine known as DAPI (5(6)amino-1-(4′-aminophenyl)-1,3,3-tri-methylindane) with commercial dianhydrides

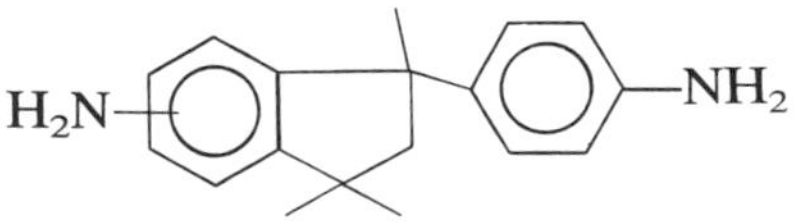

Such combinations often allow solubility in relatively non-polar solvents to be achieved. The shelf-life of a polyimide is also very dependent upon the constituent diamine and anhydride used. Polyimides derived from bis(4-amino phenyl)thioether are stable for several months in boiling water whilst those derived from metaphenylene and paraphenylene diamines rapidly become brittle under the same conditions. Some typically used amines are

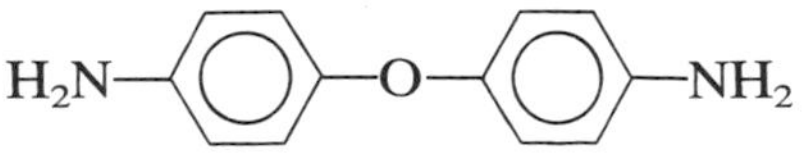

Bis(-4-aminophenyl) ether

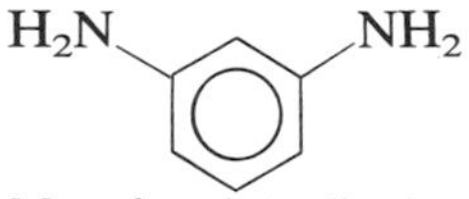

Metaphenylene diamine

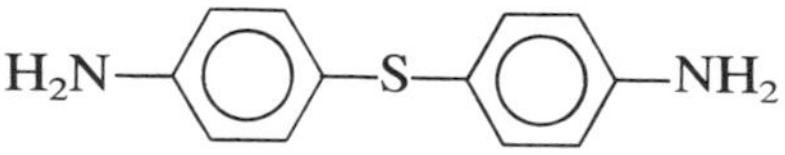

Bis(4-aminophenyl)sulphide

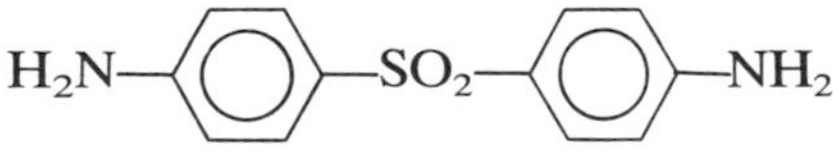

Bis(4-aminophenyl)sulphone

Polyimides are popular in the electronics industry because they have

the following outstanding properties:

(1) good thermal stability
(2) high temperature stability to oxidation
(3) flame resistance
(4) wear resistance
(5) solvent resistance
(6) radiation resistance
(7) chemical resistance.

Whilst their molecular structure gives polyimides outstanding thermal properties, under certain conditions some degradation may still occur. At temperatures below 300°C, polyimides are susceptible mainly to hydrolytic types of reaction, whilst above 300°C in air the material decomposes by a free radical initiated oxidation process. In inert atmospheres at high temperatures, pyrolysis occurs yielding carbon monoxide and carbon dioxide as the major decomposition products. The nature of any substrate materials (particularly metals) in contact with a polyimide may have a profound influence on the decomposition rate.

As a class of compounds polyimides are very resistant to organic solvents, moisture, lubricants and photoresist strippers. Resistance to acids and alkalis is not so good, although it is adequate for many applications. Their moisture absorption is also quite high, being greater than for both silicones and epoxides. Hydrolytic stability at 100°C is excellent for many polyimides with some free films being able to tolerate boiling water for a year with no deterioration.

Polyimides generally possess good adhesion and they are becoming increasingly important as die attach materials. They adhere well to aluminium and synthetic alumina surfaces, and the adhesion can be improved by using aluminium organic chelate coupling agents. Commercial aliphatic amino siloxane coupling agents also improve adhesion but thermal degradation may occur in air above 400°C. Adhesion to clean silicon dioxide surfaces is attained after annealing at 350°C.

Polyimides have found numerous applications in many areas of the electronics industry,[10,11] typical examples being:

(1) as a mask during ion implantation
(2) as an alpha particle barrier
(3) as a planarising interlayer dielectric for multilevel processes
(4) as a protective overcoat
(4) as a die attach material

Multilevel interconnection is essential for the realisation of large-scale integration in monolithic integrated circuits and polyimides offer considerable assistance in achieving this goal. Difficulties have often been experienced in bilevel aluminium structures because of the problems in depositing reliable insulating layers at low temperatures on the first aluminium layer. In the past, SiO_2 and phosphorus doped SiO_2 deposited by chemical vapour deposition (CVD) have been utilised but coverage of steps was often poor. Polyimides can be spun onto the first metallisation layer and cured to give a smooth surface. Planar multilevel structures can then be achieved with very low defect densities. A new polyimide, isoindoloquinazolinedione (PIQR), has been synthesised by Hitachi especially for this purpose. A schematic diagram of a multilayer metallisation structure utilising polyimide as a dielectric is shown in Fig. 10.2.

With the increase in packing density of dynamic RAMs, soft errors caused by alpha particles are becoming an increasingly important cause of device malfunctioning. (For further information on soft errors see Section 10.3.) Alpha particles are emitted from uranium and thorium isotopes residual in most packaging materials and thin films of polyimides deposited upon a chip surface have been successfully evaluated as alpha particle barriers. Polyimides can be made pure enough to contain no detectable amounts of uranium or thorium and a 40 μm thick film may reduce the soft error generation rate by up to 1000 times. A typical example of a commercially available polyimide alpha particle barrier is DuPont's Pyralin® PIH 61454.

Polyimides have also been used with a molybdenum mask and reactive RF sputtering as the basis for a promising new lift-off technique for producing fine metallisation patterns. The polyimide film was used in the lift-off layer enabling deposition at high temperatures and offering the following advantages:

(1) Fine patterns can be obtained, since the metallisation layer was

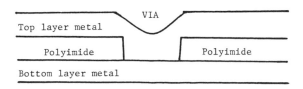

Fig. 10.2. A multilayer metallisation structure utilising polyimide as a dielectric.

deposited through sharply edged reverse masks accurately formed by sputter etching the polyimide layer.

(2) The polyimide is able to withstand temperatures of up to 350°C required for metallisation layer deposition.

(3) Electrolysis of the molybdenum mask enables the lift-off layer to be removed much more quickly than with standard lift-off techniques.

Polyimides have been successfully used as an alternative to the standard inorganic glass passivation layers used on devices. Although their moisture permeability may be considered detrimental to their use as passivation layers, other properties of polyimides are ideal. They have low permeabilities to sodium and other ionic contaminants and excellent mechanical properties. In the presence of coupling agents polyimides bond strongly to the chip surface and their low pin hole counts make them a good alternative to CVD SiO_2. When used in passivation applications it is essential that polyimides are fully cured otherwise they behave as ionic contaminants due to dipole effects.

There has been considerable interest in the use of polyimides in photoresist applications and their photosensitisation has been successfully achieved. This photosensitisation is used to crosslink the polyimide thereby increasing its molecular weight and lowering its solubility in developing solvents. Limitations of this application have so far been the low photosensitivity of the material and a resolution limit of 5–10 µm.

The high temperature stability and good wear characteristics of polyimides have been utilised in abrasion-resistant wire insulations and cables have been produced that are capable of withstanding 260°C. Another relatively new application of polyimides is in the formulation of transfer moulding compounds for electronic components to provide an alternative to epoxide materials and to give the additional advantage of high temperature stability. This moulding compound is a mineral-filled resin that is processed using similar conditions to the more conventional epoxide products. The manufacturers of this material claim that it is very low in ionic contaminants with a high glass transition temperature and a low thermal expansion coefficient. Various other applications have been found for polyimides and these are largely as electrical insulators where the material's outstanding thermal, mechanical and electrical properties can be fully utilised. As well as for wire and cable insulation, polyimide films have been used for flexible printed circuit boards and flat cable construction. Other insulation applications include coil bobbins and capacitor end caps.

The polyimide family of materials presents an example of a range of tailor-made materials whose molecular structure is designed to produce high temperature stability combined with excellent mechanical and electrical properties. The thermal stability of these polymers is related to their largely aromatic and cyclic character and the high degree of interchain polar interaction which produces a highly rigid structure. As a class of compounds the polyimides are unique in character and can thus be seen to have numerous diverse applications throughout the electronics industry which, with advances in formulation, are likely to become even more widespread.

10.6. POLYMER FILMS AS CAPACITOR DIELECTRICS

Early capacitors were usually made with paper or mica. Paper was used for large capacitance values and mica was used for high-frequency capacitors. With the advent of commercial polymer film production, these increasingly available materials became popular as capacitor dielectrics. Polymer films are ideally suited for capacitor manufacture because they:

(1) are available in large quantities at relatively low cost
(2) have high dielectric strength and are easily wound into capacitors
(3) are resistant to degradation by heat, oxygen and moisture
(4) are available with a variety of molecular structures offering a choice of dielectric, physical and chemical properties for a given application

Many polymer films have been examined as capacitor dielectrics but those principally used are polyethylene terephthalate, polystyrene, polycarbonate, polytetrafluoroethylene, polyimides and polyparaxylylene. A number of these materials are more familiar by their trade names:

Material	Trade name	Company
Polyethylene terephthalate	Mylar	DuPont
Polyimide	Kapton	DuPont
Polytetrafluoroethylene	Teflon	DuPont
Polyparaxylylene	Parylene	Union Carbide

Polyethylene terephthalate (PET) is the material used to manufacture polyester film capacitors. Its molecular structure is based on the repeat unit

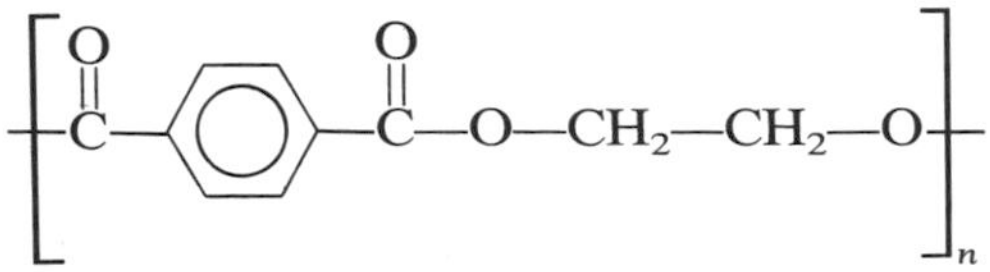

$$\left[\begin{matrix} & O & & & O & \\ & \parallel & & & \parallel & \\ \!-C\!-\!\!\bigcirc\!\!-\!C\!-\!O\!-\!CH_2\!-\!CH_2\!-\!O\!- \end{matrix} \right]_n$$

It is a tough, hard polymer with a melting point of ~265°C. PET has a very high tensile strength and modulus of elasticity which enables it to be drawn into very thin films suitable for use on capacitor winding equipment. The polar groups in the molecular structure of PET give it a reasonably high dielectric constant but its dissipation factor is acceptable for most low-frequency applications and its dielectric strength is very high.

Polystyrene (PS) is another popular capacitor dielectric and this material has a lower dielectric constant than PET because it does not contain any polar groups in its molecular structure which is based upon the repeat unit

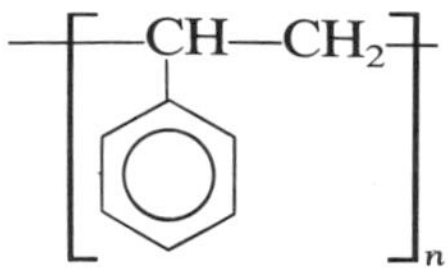

$$\left[\begin{matrix} -CH\!-\!CH_2- \\ | \\ \bigcirc \end{matrix} \right]_n$$

Polystyrene also has a low dissipation factor that, along with its dielectric constant, is virtually independent of frequency making it suitable for high-frequency applications. Some of the properties important in capacitor design are compared for polystyrene and PET in Table 10.7.

Table 10.7
Properties of Polystyrene (PS) and Polyethylene terephthalate (PET)

Property	PS	PET
Melting point (°C)	240	~265
Dielectric constant at 1 kHz	2·6	3·2
Dielectric strength (kV/mil)	5	10
Dissipation factor at 1 kHz	0·0001	0·004

Another common type of capacitor is that made with a poly-carbonate (PC). Polycarbonate is the general term used to describe the polyesters formed from the condensation reaction of a carbonic acid and a bisphenol

$$\left[O \!-\!\! \bigcirc \!\!-\! R \!-\!\! \bigcirc \!\!-\! O \!-\! \overset{\overset{\displaystyle O}{\|}}{C} \right]_n$$

These dielectric materials generally possess a good combination of electrical and physical properties making them well suited to general-purpose applications.

For the production of capacitors that must function at elevated temperatures, a dielectric of polytetrafluoroethylene (PTFE) or a polyimide (PI) may be selected.

The chemistry of polyimides has previously been reported but the repeat unit of polytetrafluoroethylene is based upon the structure

$$\left[\begin{matrix} F & F \\ | & | \\ C\!-\!C \\ | & | \\ F & F \end{matrix} \right]_n$$

PTFE has a high softening temperature and poor flow characteristics which mean that the fabrication of films for capacitors is rather difficult. The more usual method for making films thin enough for capacitor applications is to dip coat long lengths of aluminium foil with a thin film of a PTFE emulsion which is then heat-treated and sintered (see Section 10.11). Important properties of these two materials are summarised in Table 10.8.

In order to make a metallised film capacitor the polymer has a layer of either aluminium or zinc vacuum evaporated on to its surface. Once the capacitor has been wound a gradually increasing potential known

Table 10.8
Properties of Polytetrafluoroethylene (PTFE) and Polyimide (PI)

Property	PTFE	PI
Dielectric constant at 1 kHz	2·0–2·08	3·5
Dielectric strength (kV/mil)	1·0–4·5	3·0–6·0
Dissipation factor at 1 kHz	0·0002–0·0003	0·003

as the clearing voltage is applied to cause localised breakdown discharges at any weak spots in the dielectric, which in turn cause the evaporation of small localised regions of the thin metallised electrodes around the faults thus terminating the discharge.

One other polymeric material that was once used in capacitor manufacturing but which fell from favour many years ago, largely because of its toxicity and the general demise of paper capacitors, is polyvinyl carbazole. This material is worthy of mention if only because of more recent interests in its monomer's properties as a component for plasma developable resist systems.[12] Polyvinyl carbazole is a hard plastic with a structure related to polystyrene, the difference being the presence of a carbazole group in place of a phenyl group

Its main use was as an impregnant for paper capacitors as an alternative to oil impregnation. The capacitors were evacuated and impregnated with vinyl carbazole monomer. A combination of heat and pressure was then used to polymerise the monomer. Polyvinyl carbazole impregnated capacitors gave better sustained high temperature performance than the then conventional oil impregnated capacitors.

Most of the materials used as films for capacitor dielectrics are very stable compounds and offer greater reliability than could be achieved with paper capacitors. The main causes of degradation of capacitors are oxygen, moisture and heat. At the normal operating temperatures of a capacitor oxidation does not usually present a problem. Polyesters may be susceptible to hydrolysis but in typical environments the material would have to be exposed to high humidity for many years before degradation was serious. The presence of moisture may, however, affect the dielectric properties and so careful packaging is required to prevent or retard moisture absorption. All polymeric materials will have an upper operating temperature limit but under normal operating temperatures of electronic equipment temperature does not present a problem or can at least be allowed for by the use of PTFE- or PI-based capacitors.

10.7. POLYPHENYLENE SULPHIDE AS AN ENCAPSULANT FOR ELECTRONIC COMPONENTS

As has already been stated in Chapter 5, the principal method used for the plastic encapsulation of electronic devices is transfer moulding with thermosetting epoxide resin based moulding compounds. A number of other thermosetting compounds are also occasionally used but until recently no thermoplastic materials capable of being injection moulded were suitable for the encapsulation of electronic components because of the high pressures required. One thermoplastic material that has been successfully developed for the packaging of electronic components is polyphenylene sulphide.

Polyphenylene sulphide (PPS) was first produced in quantities that enabled its structure to be elucidated in 1948 but it was not until 1972 that the commercial exploitation of the material was begun by the Phillips Petroleum Company who named the product Ryton®. As produced, it is a thermoplastic white powder, with a molecular structure based upon the repeat unit

PPS is insoluble in all known solvents below 200°C but above this temperature it is partially soluble in some aromatic and heterocyclic solvents. Its chemical structure imparts excellent thermal stability and maintenance of mechanical properties at elevated temperatures. PPS is rated as non-burning by industry flammability tests.

The Ryton polyphenylene sulphide compounds are formulated for the encapsulation of semiconductors and discrete components and are particularly suited to this application because of their excellent thermal stability, resistance to chemical environments, flame retardancy and good electrical properties over a wide temperature range.[13] In order to be suitable for the encapsulation of microcircuits and transistors the material must have good flow during moulding, good dimensional stability, low levels of impurities and low moisture absorption characteristics.

Just as with the thermosetting materials the melt viscosity and flow of PPS are determined by the filler content and the molecular weight of the polymer. Similarly the normally high thermal expansion coefficient associated with thermoplastic materials can be reduced by the careful selection and quantity of fillers. The formulated moulding compound is made from selected high-purity materials in order to minimise the levels of ionic contaminants and the overall moisture permeability is low, largely because of the molecular structure.

The encapsulation of electronic components in a thermoplastic material by injection moulding can offer distinct advantages over conventional transfer moulding with thermosetting materials. The use of a thermoplastic material eliminates the cure time required with thermosets because the compound only has to be melted, injected into the mould cavities and cooled below its melting point. A fully automated process using a thermoplastic material could conceivably have a cycle time of less than 15 s, which is approximately one third of the fastest cycle times currently achievable with automated transfer moulding of epoxides. Thermoplastic materials such as polyphenylene sulphide also do not flash as thermosetting materials do and thus there is a reduction in the time required for mould cleaning. The other main advantage of using a thermoplastic material is that the runners and cull of each shot can be reground and used again so there is a reduction in material wastage.

Polyphenylene sulphide has been used commercially to encapsulate ceramic capacitors for several years and mica capacitors encapsulated in Phillips Ryton BR61B have recently received the British Standard 9000 approval (BS9070-N-002). More recently, attempts have been made to encapsulate wire-bonded integrated circuits, which is one of the most difficult applications to successfully mould with a thermoplastic material because of problems with wire-bond sweep. By paying particular attention to the melt viscosity of the PPS and having careful control of the moulding parameters it has been shown that it is possible to mould successfully both TTL and linear operational amplifier type devices.

Polyphenylene sulphide's low dielectric constant and dissipation factor coupled with a high dielectric strength make both the filled and unfilled materials useful in a variety of applications as electrical insulators. It has been successfully employed in the manufacture of connectors, coil bobbins and pin cushion connector coil terminal supports for colour televisions.

Polyphenylene sulphides are a range of compounds offering a unique combination of mouldability, good electrical properties, solvent and chemical resistances, thermal performance and high flexural modulus which will ensure their continued use in the electronics and electrical industries.

10.8. POLYURETHANES

Polyurethanes represent a class of compounds that are used extensively as coatings, largely because they have a combination of properties making them ideal for this type of application. Their properties depend upon the formulation and they can be made as both one- and two-part systems exhibiting outstanding electrical insulation, abrasion resistance and toughness coupled with flexibility, moisture and chemical resistance.

Polyurethanes are formed by the reaction of diisocyanates or polyisocyanates with polyester or polyether resins. Two of the most widely used isocyanates are tolylene diisocyanate (TDI) and diphenyl methane diisocyanate (MDI)

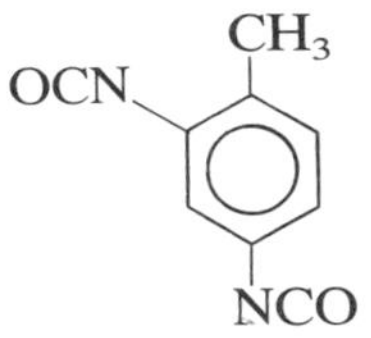

2,4-Tolylene diisocyanate 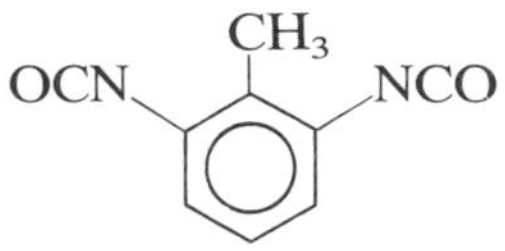   2,6-Tolylene diisocyanate

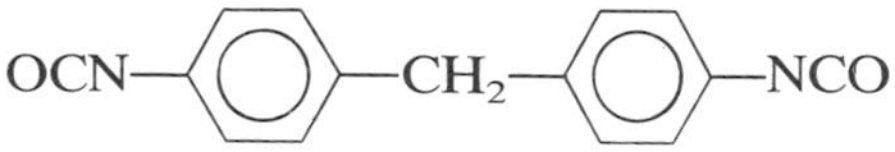

Diphenyl methane diisocyanate

These materials tend to be rather toxic and possess high vapour pressures which make them difficult to use. In order to lower the vapour pressures and make them safer to handle they are usually modified by co-reacting all or a large proportion of the isocyanate groups. A typical example would be to react the isocyanates with trimethyl propane, which reacts with three available isocyanate groups from three isocyanate molecules yielding a low vapour pressure adduct, for example

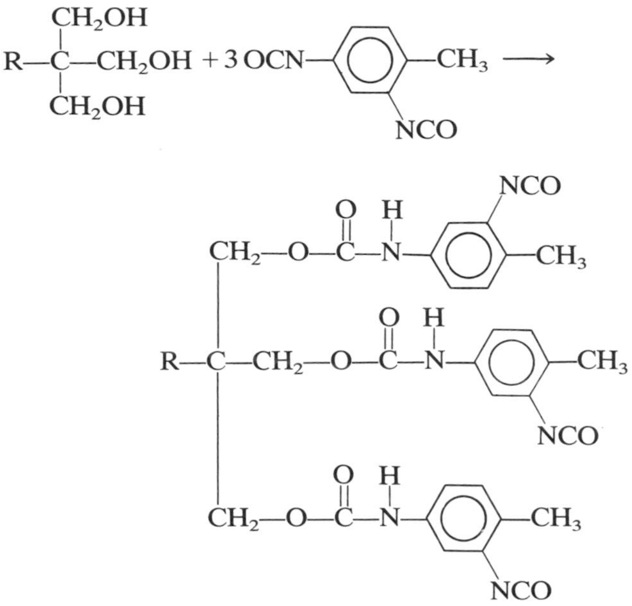

$$R—\underset{\underset{CH_2OH}{|}}{\overset{\overset{CH_2OH}{|}}{C}}—CH_2OH \; + 3\, OCN—\langle\bigcirc\rangle\underset{NCO}{—}CH_3 \; \longrightarrow$$

The basic reaction used in the polymerisation reaction to form polyurethanes is the addition of an hydroxyl group to an isocyanate group to give a urethane linkage

$$\overset{\delta^-\;\;\;\delta^+}{R—N}\!\!\doteq\!\!C\!\!=\!\!O \; + \; \overset{\delta^+\delta^-}{HO—R'} \; \longrightarrow \; R—\overset{H}{\underset{|}{N}}—\overset{O}{\overset{||}{C}}—OR'$$

Other groups containing active hydrogen atoms (such as amino and carboxyl groups) are also able to react with the isocyanate group but these lead to the formation of urea- and amide-type linkages.

Urethanes can be cured by any one of five different methods and may be formulated in the following ways:

(1) as a single-component pre-reacted system
(2) as a single-component moisture-cured system
(3) as a single-component heat-cured system
(4) as a two-part system cured either catalytically or by reaction with a compound having active hydrogens

In the single-component pre-reacted type of formulation a polyiso-cyanate has been pre-reacted with a polyhydric alcohol ester of a fatty acid. These compounds are referred to as uralkyds, urethane oils or oil-modified urethanes and their main attraction is their low cost, which is largely due to the use of naturally occurring oils such as castor oil. They have good moisture and alkali resistances, but their overall performance characteristics are not as good as some of the other polyurethanes in terms of suitability for electrical applications.

The moisture-cured polyurethane systems consist of an isocyanate-terminated resin or prepolymer adduct which is usually applied in solution as a coating. The solvent evaporates leaving a thin film that reacts with moisture in the air to undergo a crosslinking reaction via the formation of urea linkages and the evolution of carbon dioxide. The carbon dioxide is not generated rapidly enough to form bubbles and slowly diffuses into the air as it forms leaving behind a clear film. A typical formulation for a moisture-curing prepolymer resin would be:

Component	Moles
Tolylene diisocyanate	8
Trimethyl propane	2
1,3-Butylene glycol	1
Polypropylene glycol	1

To give a usable product other ingredients such as solvents, flow control agents and antioxidants would also be added to the formulation.

The one-component thermally cured isocyanates are the ones most commonly used for electronic and electrical applications finding particular use as wire coatings and enamels. In order to make the isocyanate latent until heated, these reactive groups are blocked by a highly volatile or easily decomposed compound. Typical examples are malonic and aceto acetic esters and phenol, which are the most popular materials for commercial use. When heated above 160°C the phenol is released regenerating the isocyanate groups. If this temperature is too high for a given application then blocking of the isocyanate with a malonic ester will give a lower unblocking temperature.

The two-component catalytically cured polyurethanes are comprised of one part containing the isocyanate and a second part containing a catalyst such as a polyamine or a cobalt-based salt. Once the two components are mixed the system will have a limited pot life dependent upon the type of catalyst and quantity used.

Table 10.9

Curing Reactions and Properties of Various Polyurethanes

Type	Cure reaction	Free isocyanate	Pot life	Order of preference for electronics applications
One-part pre-reacted	Air	No	Unlimited	4
One-part moisture-cured	H_2O	Yes	Extended	5
One-part heat-cured	Heat	No	Unlimited	1
Two-part catalyst-cured	Amine	Yes	Limited	3
Two-part polyol-cured	NCO/OH	Yes	Limited	2

The final type of polyurethane available is known as the two-part polyol system. One component contains the isocyanate containing compound such as a prepolymer or adduct and the second component is a hydroxy group terminated resin which may or may not contain a catalyst. The most commonly used hydroxy-terminated components are polyols, castor oil, hydroxy-terminated polyesters and some epoxy resins. By varying the components of such a system a wide range of cured properties can be achieved ranging from high flexibility to very hard or brittle. The curing reactions and properties of the different polyurethanes are summarised in Table 10.9.

Polyurethanes offer a combination of properties which make them ideal as coating materials for electrical and electronic applications. With the correct formulation they are capable of giving films that exhibit excellent adhesion, high gloss, water and solvent resistances, low gas and moisture permeabilities and outstanding electrical properties.[14] Formulations are also available which produce solvent removable linear polyurethane encapsulants allowing a rework capability for expensive electronic packages.[15]

10.9. SILANE COUPLING AGENTS

Although the silane coupling agents are not polymeric materials as such, they are used extensively in conjunction with polymeric adhesives, coatings and mineral filled moulding compound formulations to enhance their overall properties. They are worthy of mention not only because of this fact but also because they actually react with the polymer matrix becoming chemically linked to it and thus forming part of the cured polymer network. Many polymer applications would not be possible without the use of these extremely valuable compounds and this section attempts to highlight some of the properties which make them particularly useful in conjunction with plastics for electronics.

Silane coupling agents are a unique range of compounds which can effectively overcome the problems encountered when components of different chemical nature and physical form are in intimate contact. An example would be the interfacial region between the organic epoxide resin and the inorganic filler particles of an epoxide moulding compound. Because of their unique molecular structures, silane coupling agents are often able to effectively eliminate these interfaces by forming bonds with both the organic and the inorganic components.

As their name implies, silanes are based upon silicon and they can be represented by the general formula

$$XRSi(OR')_3$$

where X represents an organic functional group, R represents the chemical linkage between the reactive functional group and the silicon and the $Si(OR')_3$ part provides the ability to react with inorganic materials such as filler particles. Thus we have a linear molecule which has a reactive group at one end capable of being incorporated into the matrix of a curing polymer and a silicon-based group at the other end which bonds to inorganic materials.

The action of a silane in bonding two dissimilar materials together can be represented schematically in three stages. The initial reaction involves hydrolysis of the inorganic portion of the coupling agent to a silanol group by moisture

$$XRSi(OR')_3 \xrightarrow{\text{H}_2\text{O}} XRSi(OH)_3$$

This hydrolysis product is then able to form a chemical bond with the filler surface via surface OH groups

$$XRSi(OH)_3 + HO\!-\!\boxed{\text{Filler}} \longrightarrow XR\overset{|}{\underset{|}{Si}}\!-\!O\!-\!\boxed{\text{Filler}}$$

The silane-treated filler is then mixed in with the other ingredients of the formulation and the organofunctional (X) groups are able to react with and bond to the resinous component when it cures

$$\text{Reactive resin} + \text{XRSi} \underset{\overset{|}{O}}{\overset{\overset{|}{O}}{|}}\!\!-\!\!O\!\!-\!\boxed{\text{Filler}}$$

$$\longrightarrow \text{Reactive resin}\!-\!\text{X}\!-\!\text{R}\!-\!\underset{\overset{|}{O}}{\overset{\overset{O}{|}}{Si}}\!-\!\boxed{\text{Filler}}$$

The main purpose of the coupling agent is to effectively eliminate the interface between the two components. This allows the maximum adhesion to be achieved between a filler and a resinous component and stress transfer can occur from the low strength resin to the higher strength filler. In addition to the stress transfer characteristics and good adhesion, the use of a silane coupling agent also allows these properties to be maintained during exposure to harsh environments such as high temperature and humidity. The electrical properties of a silane-treated material will also be better than those of the untreated material with dielectric constants, dissipation factors and volume resistivities being stabilised.

Silane coupling agents are available with a variety of organo-functional group terminations making them compatible with a large number of different polymers. Some of the representative organo-functional silanes that are currently available in commercial quantities are shown in Table 10.10.

Whilst the choice of silane coupling agent obviously plays a large part in determining the degree of improvement achieved, the method of treatment is also particularly important. For the treatment of particulate fillers, as might be the case for the formulation of an epoxide resin based transfer moulding compound, there are essentially two approaches that can be adopted. These are pretreatment of the filler and integral blending. The pretreatment of the filler, as its name implies, involves pretreating the filler prior to its incorporation into the bulk formulation of the compound. In this process the silane coupling agent is diluted with a suitable solvent such as alcohol/water and the solution is tumble mixed with the dry filler. The treated filler is then heat dried to remove the solvent and water. Although the amount

Table 10.10
Silane Coupling Agents: Structure and Applications

Compound name	Structure	Formula molecular weight	Polymers for use with
Vinyl triethoxy-silane	$CH_2{=}CHSi(OC_2H_5)_3$	190·3	Unsaturated polyesters, polyolefins
Aminopropyl-triethoxysilane	$H_2N(CH_2)_3(SiOC_2H_5)_3$	221·3	Epoxides, phenolics, polyesters
Glycidoxy-propyltri-methoxysilane	$CH_2{-}CHCH_2O(CH_2)_3Si(OCH_3)_3$ $\diagdown_{O}\diagup$	236·1	Epoxides, polyesters
Mercapto-propyltrimethoxy-silane	$HS(CH_2)_3Si(OCH_3)_3$	196·2	Epoxides, poly-sulphide

of silane coupling agent required for a given inorganic filler will depend upon such factors as its type and particle size, a typical starting value to use as a guideline would be ~1%. The integral blending method simply involves the addition of a silane coupling agent to the blend of materials used for a given formulation. This method has the advantage of avoiding the separate pretreatment operation but may require more material to be used in order to ensure optimum coverage of the filler surface. Quite often it is possible to buy ready treated fillers; commercially treated silica, silicates, clays, micas, mineral fibres and alumina trihydrate are readily available. Glass fibres, as might be used in printed circuit board formulations, are also available ready treated.

Silane coupling agents find extensive use in all types of thermoplastic and thermosetting materials where components with improved shear strength, chemical resistance and electrical properties are required. They are used with mineral filled epoxides to provide continued electrical insulation and low loss properties after extensive exposure to moisture. With thermoplastic materials, silane treatment of fillers is thought to protect against mechanical damage during high shear operations such as mixing, extrusion and injection moulding. Filled polyethylene to be extruded as cable coatings is modified with silanes

Table 10.11
The Effects of Silane Treatment upon the Flexural Strength and Electrical
Properties of a Quartz-Filled Anhydride Cured Epoxide[16]

	Control (no silane)	Silane and loading (wt %)[a]		
		1 0·31	2 0·30	3 0·28
Flexural strength (psi)				
Dry, room temperature	21 000			20 000
72-h boil	9000			14 000
Dielectric constant				
Dry	3·39	3·48	3·40	3·45
72-h boil	14·60	3·52	3·44	3·47
Dissipation factor				
Dry	0·017	0·016	0·016	0·013
72-h boil	0·305	0·023	0·024	0·023
Volume resistivity (ohm-cm)				
Dry				
72-h boil				
Dielectric strength (V/mil)				
Dry	381	367	357	355
72-h boil	103	360	391	355

[a] Percent by weight loading based on weight of quartz filler. 1 = (3,4 epoxy-cyclohexyl)ethyl trimethoxy silane; 2 = glicydoxypropyl trimethoxy silane; 3 = aminopropyltriethoxysilane.

to improve the wet electrical properties of the material. The improvements in flexural strength and electrical properties achieved by silane treating the quartz fillers used in an anhydride-cured epoxide formulation are shown in Table 10.11.

Another important application for silane coupling agents is as an adhesion promoter for polyimides and photoresists on silicon wafers during processing operations. Where polyimides are used as interlayer dielectrics in multilayer metallisation deposition, the use of a silane coupling agent such as aminopropyltriethoxysilane is very effective in preventing stripping of the thin films.

Silane coupling agents are thus an extremely valuable range of compounds that can effectively and economically improve the adhesive bonding of a polymer to a mineral, giving superior hydrolytic stability and enhanced electrical and mechanical properties in composite materials.

10.10. DIE ATTACH MATERIALS

When semiconductor devices were largely packaged in ceramics the usual method of die attach was to use a silicon–gold eutectic bond. The relatively high melting point of this eutectic presented no problems to the ceramic pack and higher temperatures were often used to seal the lids onto the packages. Polymer-based adhesives did exist but they were unable to stand the high temperatures used with ceramic and glass packages and only began to be seriously considered when plastic encapsulation using transfer moulding techniques became more popular. These initial compounds were epoxide resin based and whilst they may have been suitable for use with low-cost plastic encapsulated consumer electronics there were serious doubts about their use as die attach materials with hermetically sealed hybrids. The main areas of concern here were with outgassing, chemical composition and the possibility of contamination of the die.

The use of polymer-based adhesives for semiconductor die attachment offers a number of advantages over the more traditional eutectic die bond approach. Polymeric die attach materials can be used at lower temperatures (150°C or less), have re-work capability and, because of their low moduli of elasticity, allow stress dissipation without damaging the die or the substrate. This latter factor has become particularly important in the last few years with the ever-increasing size of die.[17,18]

Die attach materials are often required to be thermally and/or electrically conducting. Thermal conductivity is achieved relatively easily by the use of alumina as a filler (e.g., Epotek H61). In order to achieve electrical conductivity, metallic fillers are required and the usual components are gold or silver. The most popular of these two metals is silver since it is considerably cheaper, although gold-filled polymer adhesives are used for ultra high reliability applications. Filler loadings in the region of 60 to 75% by weight are normally required to give the required conductivity. The properties of some commercially available die attach materials are shown in Table 10.12.

Epoxide die attach materials are available in a wide variety of forms but the main types are two-part pastes, one-part pastes, unsupported films and supported films. The principal two types are the one- and two-part pastes. A variety of possible epoxide chemistry systems is available for both one- and two-part cured materials. A typical two-part formulation would be based on a diglycidyl ether of bisphenol A or bisphenol F and be cured with an aliphatic amine or polyamide.

Table 10.12

Properties of Commercially Available Silver-Filled Epoxides[a]

Material	System type	Recommended cure schedule	Volume resistivity (Ω cm)	Shelf-life	Tensile shear strength (lb/in^2)
Epotek H20E	2 Part	45 s @175°C 5 min @150°C 15 min @120°C	$1-4 \times 10^{-4}$	1 year @25°C	1500
Du Pont 4621	1 Part	1 h @175°C	$1 \cdot 3 \times 10^{-4}$	3 months @25°C	>1500
Ablebond 84–1 LMI	1 Part	1 h @150°C	2×10^{-4}	2 weeks @25°C	2000
Ablebond 36–2	1 Part	1 h @125°C 30 min @150°C	$0 \cdot 6-1 \times 10^{-4}$	1 week @25°C	1200–1800
Amicon C850–6	1 Part	1 h @150°C	1×10^{-4}	3 months @25°C	>1500
Amicon C868–1	1 Part	1 h @170°C	1×10^{-3}	3 weeks @25°C	>1500

[a] Information from manufacturers' data sheets.

For the one-part systems diglycidyl ethers of bisphenol and diglycidyl ethers of resorcinol have been utilised with hardeners such as dicyandiamide, aromatic substituted ureas and boron trifluoride amine complexes. For further information on component mounting with epoxides the reader is referred to an excellent review by Marshall.[19]

Once epoxide die attach materials became accepted for plastic encapsulated devices interest turned back to existing products that were assembled in ceramic packages but unfortunately the thermal stability of the epoxide-based adhesives was not sufficient to withstand the sealing temperatures required with sidebraze and cerdip packages. Consequently much attention has been given to silver-filled polyimide die attach materials in the hope that these high temperature polymers will have sufficient stability to withstand these elevated sealing temperatures. At the present time polyimide die attach techniques are compatible with sidebraze type packaging, where the sealing temperature is around 340°C but are not yet stable enough for use with cerdip packages which are sealed in air at temperatures of 420–450°C.

The main causes of the thermal stability problems are thought to be due to the catalytic activity of the surface of the finely divided silver particles, which cause adhesive strengths to degrade and the buildup of moisture and carbon dioxide within the package cavity. There is currently no available conductive adhesive system that is capable of withstanding the cerdip sealing process without suffering unacceptable chemical and electrical degradations.[20] Polyimide die attach materials are, however, becoming increasingly popular in standard plastic packaged die attach applications particularly where power devices or devices operating at elevated temperatures are concerned and they offer a viable alternative to epoxides.[21]

10.11. FLUOROCARBONS

The normal characteristics associated with high temperature polymers are the presence of a predominantly aromatic molecular structure and a large degree of crosslinking. High temperature polymers are thus usually thermosetting materials and thermoplastics are not normally regarded as being suitable for these types of application. There are, however, two major thermoplastics that have good thermal stability and these are both fluorocarbons: polytetrafluoroethylene (PTFE) and polychlorotrifluoroethylene (PCFE). A slightly more thermoplastic analogue of PTFE also exists and this is a fluorinated ethylene–propylene copolymer (PFEP). The basic repeat units of the molecular structures of these materials are

The fluorocarbon polymers have a number of outstanding properties which make them useful in the electronics and electrical industries:

(1) high thermal stability with continuous service at 200–260°C being possible
(2) maintained excellent electrical and dielectric properties up to 260°C
(3) high resistance to chemicals and solvents
(4) very high purity
(5) very low coefficient of friction and surface energy

All of these properties can be attributed to the molecular structure of the fluorocarbon polymers. The molecules consist of long chains of carbon atoms totally surrounded by highly electronegative fluorine atoms. The carbon–fluorine bond strength is particularly high and this, along with the electronegativity of the fluorine atoms which tends to stiffen the polymer chains, accounts for both their high chemical and thermal stabilities. These unique properties have meant that the fluorocarbons have been used in many areas of the electronics and electrical industries. Typical uses have included tape and wire insulations, particularly in high temperature, high frequency applications, capacitor dielectrics, coaxial connectors, coil wrappers, transformers and as a glass fibre laminate for printed circuit boards.

The PTFE and PFEP polymers have some of the best electrical properties of all known polymers and their more important electrical properties are shown in Table 10.13.

These excellent electrical properties are also maintained over a wide temperature range with the dielectric constant of PTFE remaining virtually unchanged up to 260°C over the frequency range 60 to

Table 10.13
Electrical Properties of PTFE, PFEP and Plasticised PCFE

	PTFE	*PFEP*	*PCFE*
Dielectric constant (1 kHz)	2·0–2·2	2·1	2·68
Power factor	$<3 \times 10^{-4}$	2×10^{-4}	$1·3 \times 10^{2}$
Volume resistivity (Ω cm)	$>10^{18}$	$>10^{18}$	$1·1 \times 10^{18}$
Dielectric strength (short time, V/mil)	3000–4500	—	—

3×10^{9} Hz. Similarly, the volume resistivity shows little variation with temperature, remaining virtually constant up to 220°C and even after prolonged exposure to moisture the value remains greater than 10^{15} ohm cm.

PTFE and PFEP are available for coating applications as colloidal aqueous dispersions which are essentially suspensions of hydrophobic negatively charged particles in water. These coatings can be applied by spraying, dipping, flowing or casting and in order to obtain a continuous uniform coating an initial bake at around 120°C is used to remove the water. The fluoropolymer is then heated to between 360 and 400°C to sinter the individual particles together. In addition to this method of application, the fluoro polymers can also be applied by extrusion, electrostatic spraying and fluidised bed techniques. One particular area in which PFEP has a distinct advantage over PTFE is in processing, since it has a melt flow comparable to most commonly used thermoplastic resins and can be processed by conventional extrusion and injection moulding equipment.

10.12. POLYXYLYLENES

Polyxylylenes are linear polymers derived from *p*-xylene or a substituted *p*-xylene and have the repeat unit

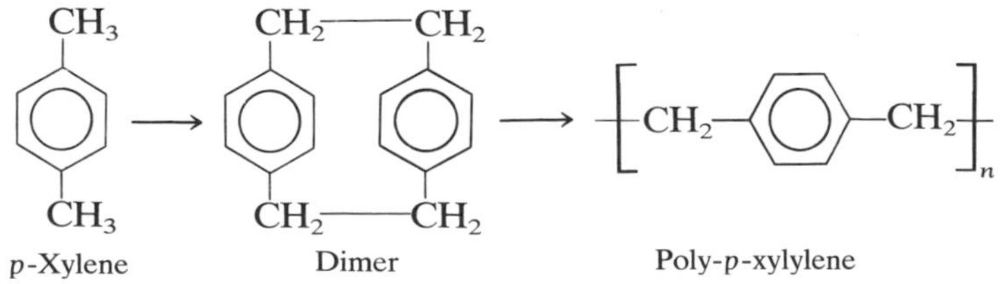

p-Xylene   Dimer   Poly-p-xylylene

Union Carbide has developed a process whereby poly-p-xylylenes can be produced which are completely linear non-crosslinked polymers. This process involves the conversion of p-xylene to a white dimeric solid known as di-p-xylylene and then pyrolysing this dimer at about 650°C in a vacuum of 0·1–0·5 Torr. The essential feature of this process is that the dimer dissociates into a monomer, p-xylylene, which immediately polymerises on contact with a surface forming a coating.[22] These polymer films are called parylenes by Union Carbide, who have a series of patents covering the chemistry of the materials and the apparatus for their vapour deposition.

Parylene coatings are unique in their method of application and were the first materials to be vapour phase depositable without the use of high surface temperatures, high vacuums or ultra-violet radiation. They are ideal for electronics coatings applications because, even though the dimer decomposition temperature is between 650 and 680°C, the piecepart to be coated can be maintained anywhere between room temperature and 150°C.

Parylene films have excellent electrical properties with high volume and surface resistivitities whilst their dielectric constants and dissipation factors are quite low, although not as good as PTFE. The two main types of parylene are Parylene N and Parylene C. Parylene N is the unsubstituted polymer, whilst Parylene C is chlorine substituted. The best overall electrical properties are exhibited by Parylene N but Parylene C is superior with respect to D.C. dielectric breakdown voltage for films under 5 μm in thickness. The principal electrical properties of parylenes are shown in Table 10.14.

Parylene films generally have low coefficients of friction, very high abrasion resistance and low moisture vapour permeabilities. They do

Table 10.14
Electrical Properties of Parylenes[a]

Property	Method	Parylene C	Parylene N
Dielectric constant (1 kHz)	ASTM D 150–65T	3·10	2·65
Dissipation factor (1 kHz)	ASTM D 150–65T	0·019	0·0002
Volume resistivity (Ω cm)	ASTM D 257–61	$8\cdot8\times10^{16}$	$1\cdot4\times10^{17}$
Surface resistivity (Ω)	ASTM D 257–61	10^{14}	10^{13}

[a] Union Carbide Corporation data.

not have good temperature stability in air, however, largely because of the presence of the aliphatic ethylene group adjacent to the benzene rings, which allows oxidation and thermal cleavage to occur.

Parylenes have found a number of applications in the electrical and electronics industries, the major one being the conformal protective coating of printed circuit boards. They are particularly good at covering closely packed boards since they have the ability to penetrate under and around the various components. In addition to excellent conformal coating characteristics the parylene's high electrical resistance and stability during humidity and temperature cycling makes them particularly useful as a barrier coating.

Parylenes have also been used as an additional protective layer to the inorganic glass passivations used in semiconductor device fabrication. The main advantage of this extra coating is its ability to cover the normally exposed bond pad areas, giving protection to the exposed aluminium. Similar applications can also be envisaged for the coating of hybrid assemblies.

10.13. POLYMERS FOR OPTICAL DATA STORAGE MEDIA

With the ever-increasing need to store and retrieve large amounts of data, there is a growing interest in economical, compact, high-speed storage systems. Polymeric materials are already used as the basis of magnetic tape recording data storage methods but further improvements are becoming much more difficult to achieve. One promising new area that may meet the future requirements for data storage is that encompassing optical technologies. A number of systems have been proposed but one of the most attractive methods currently available utilises a highly focussed laser beam for optically recording data onto a disc format at very high rates with extremely high packing density. Playback-only systems have been available for some years, but the greatest challenge is with record/playback systems and it is the recording media which represent the key components of such a system.

Many different optical recording media have been investigated and among these polycrystalline tellurium films have several attractive properties. Unfortunately they also show rapid degradation in high humidity or oxygen environments. Like many metals tellurium is oxidised upon exposure to air, but under dry conditions a stable passivating layer is rapidly formed preventing further oxidation. Water

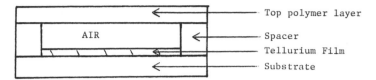

Fig. 10.3. The air sandwich structure used with tellurium-based optical data storage systems.

destabilises this thin oxide layer and oxidation of the tellurium metal proceeds until it is all consumed. The opportunity for polymeric materials in these types of application is as a barrier or protecting layer to protect the thin layers of tellurium from environmental gases and moisture likely to cause degradation. Smith *et al.*[23] examined the feasibility of protecting thin metal layers from oxidation in the light of known vapour permeability characteristics of selected organic polymeric materials.

Attempts have been made to increase the tellurium stability by using various polymers as overcoats and to form sandwich-type structures. The air sandwich structure utilises a substrate on which the tellurium metal is deposited. Annular spacers then provide the supports for the top polymer film and leave a cavity immediately above the metal, as shown in Fig. 10.3.

Overcoated or encapsulated structures have also been investigated, as shown in Fig. 10.4, but only a few materials can be utilised without the metal film suffering a significant loss in marking sensitivity.

The concept of preventing metal oxidation by encapsulation is very much the same as previous examples cited for other electronic applications in that if water and oxygen can be prevented from reaching the metal surface no oxidation will occur. Since all polymers are permeable to moisture and gases the problem is one of kinetics, i.e. for how long oxidation of the metal will be insignificant enough not to affect the marking or readout sensitivity. It is thus important that the polymeric

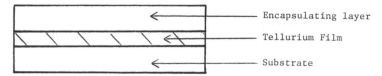

Fig. 10.4. The overcoated structure used with tellurium-based optical data storage systems.

coating materials selected have the optimum barrier properties to the detrimental species.

Recent work carried out by Plessey has taken the concept of optical read–write data storage systems one step further. This has involved the use of a range of novel organic photochromic materials as the basis for a reversible optical data storage system. The main reason for the development of such a system was the hope that optical methods would increase the power of computers by allowing parallel processing, giving very high data rates together with a greatly increased storage capacity. The low cost of optical data storage resulted from the use of a high-resolution storage medium. Photographic emulsions are capable of recording in excess of 10^6 bits of data/mm^2, with the result that the contents of the Concise Oxford Dictionary could be contained on a strip of 35-mm film only a few tens of millimetres long. The system developed by Plessey utilised holographic recording techniques for storing images on photosensitive materials. For reversible storage a photochromic film was used with the recording operation carried out by first exposing with ultra-violet light to impart a colour to the film and then recording by bleaching areas of the colour with green light. Erasure was achieved by re-exposure to the ultra-violet light and data readout utilised a weak green light source of sufficiently low power to cause no further bleaching.

The organic photochromic materials were coated onto both cellulose acetate and polyethylene terephthalate film bases. Unfortunately most organic photochromic materials eventually suffer from fatigue problems which limit the achievable number of read–write cycles. This has been directly related to the presence of oxygen which has permeated the coating material causing a decrease in the photochromic response by the formation of unwanted side products. Wilson[24] examined the photo-induced fatigue of these photochromics in various polymeric matrices and correlated this with the oxygen permeability of the polymers. Figure 10.5 clearly shows that the rate of fatigue is greatest for the photochromic in the polymer matrix with the highest oxygen permeability.

Material developments have now produced photochromics with greater fatigue resistance and with the correct selection of film base materials and the judicious use of antioxidants the problems of oxygen-induced fatigue have been significantly reduced.

With the further development of new photo-erasable materials and advances in polymer film base technology a successful future for optical data storage seems assured.

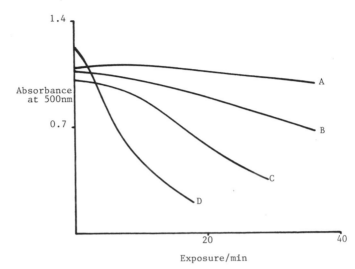

Fig. 10.5. Photochromic fatigue in polymer matrices with differing oxygen permeabilities. Relative oxygen permeability: A, 1·0; B, 6·7; C, 33·3; D, 233·3.

10.14. CONCLUDING REMARKS

This work illustrates the properties of some of the vast number of plastic/polymeric materials currently available and their applications in the electronics and electrical industries. Most of these applications have utilised plastics because of their excellent electrical, physical and mechanical properties and the majority of uses find the plastic as an insulator/dielectric rather than as the active components of electronic devices.

More recently, interest has been centred on molecular conductors as an exciting new group of electronic materials. In these materials the structure is composed of molecules linked together rather than atoms as is the case with silicon. These molecular conductors are normally insulators but the main area of interest is with conducting polymers in which electrons can move along the carbon–carbon backbone of the polymer molecule forming an almost one-dimensional conductor. Typical examples of these materials are polyacetylene, polydiacetylene, polyparaphenylene and the polypyrroles and they offer the promise of polymer-based semiconductors which will have many advantages over silicon-based devices. Polyacetylene and polypyrrole have already been used for building lightweight batteries and polypyrrole has been examined as a material for making electronic displays.

In addition to the development of these novel materials the number of possible applications for more standard plastic/polymeric materials will continue to grow and the future for plastics in electronics is certainly an exciting one. With doped polymeric materials actually forming the active components of many semiconductor devices, it may not be too long before polymer-based semiconductors, solar cells and batteries are in routine production and the all-plastic transistor radio becomes a reality.

REFERENCES

1. Shinohara, M. M., *Proc. Plastics in Telecommunications III*, IEE, London, 1982.
2. Abshier, C. S., Berry, J. and Maget, H. J. R., *Insulation/Circuits* (Oct. 1977), 27.
3. Creed, K. E., *CTBN Modified Epoxy Resin, Neutron Devices for US Energy*, R & D Administration Contract E (29-2)-656, August 1975.
4. Hill, J. W., SPE-ANTEC, San Francisco, May, 1974.
5. Bouldin, D. P., *J. Electron Materials*, **10**(4) (1981) 747.
6. Goosey, M. T., *Proc. Semiconductor 83 International*, Birmingham, September 27–29, 1983.
7. Jones, P. C., *Proc. Internepcon UK 80*, Brighton, October 14–16, 1980.
8. Meyer, C. L., *SPE Journal*, **25** (1969) 65.
9. Guillbert, C. R., US Patent 4,267,300, May 12th, 1981.
10. Iscoff, R., *Semiconductor International* (October 1984) 116.
11. Lee, Y. K. and Craig, J. D., In: *Polymer Materials for Electronic Applications*. (Edited by E. D. Feit and C. W. Wilkins) ACS Symposium Series 184, ACS, Washington, D.C., 1982.
12. Taylor, G. N., Wolf, T. M. and Goldrich, M. R., *J. Electrochem. Soc.*, **128**(2) (1981) 361.
13. Bocke, P. J., Leland, E., Jr. and DeCallatay, G., *Proc. Semiconductor 83 International*, Birmingham, September 27–29, 1983.
14. Schwartz, H. R. and Lucy, J. M., *Proc. 27th National Sampe Symposium*, May 4–6, 1982, p. 886.
15. Wischmann, K. B., *Energy Res. Abstr.*, **4**(19) (1979) Abst. No. 48951 (*C.A.*, **92,** 23715a).
16. Marsden, J. G., *Plastics Compounding* (July/Aug. 1978).
17. Kulesza, F. W., *SPE 31st Antec*, Montreal, May 7–10, 1973.
18. David, R. F. S., *Solid State Technology* (Sept. 1975) 40.
19. Marshall, C., *Electrocomponent Science and Technology*, **5** (1978) 171.
20. Moghadam, F. K., *Solid State Technology* (Jan. 1984) 149.

21. Estes, R. H. and Kulesza, F. W., *Int. J. Hybrid Microelectron.*, **5**(2) (1982) 336.
22. Kramer, P., Sharma, A. K., Hennecke, E. E. and Yasuda, H., *J. Polym. Sci. Polym. Chem. Ed.*, **22**(2) (Feb. 1984) 493.
23. Smith, T. W., Johnson, G. E., Ward, A. T. and Luca, D. J., *Proc. SPIE Int. Soc. Opt. Eng.*, *USA*, *329* (Optical Disc Technology) (1982), p. 236.
24: Wilson, A. E. J., Plessey Research, Caswell, private communication, 1983.

Appendix

EXAMPLES OF COMMERCIALLY AVAILABLE MATERIALS, TRADE NAMES AND MANUFACTURERS

Type	Trade name	Features	Application	Manufacturer
1. Adhesives				
One-part silicone	Abelbond 190-3	Silver-filled, conductive	R.F. shielding	Abelstik Labs
One-part epoxide	Abelbond 36-2	Silver-filled, conductive	Die attach	Abelstik Labs
One-part epoxide	Abelbond 41-1	Electrically insulating	Die attach	Abelstik Labs
One-part polyamide	Abelbond 71-1	Silver-filled, conductive	Die attach	Abelstik Labs
One-part epoxide	E-Solder 3205	Gold-filled, conductive	Die attach	Acme Chemicals and Insulation Company
One-part epoxide	Uniset A359	Aluminium-filled, conductive	Magnets and speakers	Amicon Corporation
One-part epoxide	Uniset C409	Silver-filled	Die attach, capacitors, resistors	Amicon Corporation
One-part silicone	Cho-Bond 1030	Conductive, resilient, flexible	Emigaskets	Chomerics
One-part epoxide	Eccobond 60C	Carbon graphite filled	Electrical connections	Emerson & Cuming
One-part epoxide	Epo-Tek H41	Gold-filled, conductive	Military and aerospace applications	Epon Technology
Two-part polyvinyl acetate	E-Solder 3044	Silver-filled, conductive	Die attach for high temperatures	Acme Chemicals and Insulation Company
Two-part synthetic	Bostik 7008	Solvent resistance	Coil winding, coatings	Bostik
Two-part polyurethane	Castak CU-2009R1	Good insulation and hydrolytic stability	Laminating, coating, potting	Castall, Inc.
Two-part epoxide	Devcon EK-80 Kit	Shock resistance	Potting, encapsulant, printed circuit	Davcon Corporation
Two-part epoxide	FRI ECR-4200	Silver-filled	Microwave, RFI shielding	Formulated Resins, Inc.
Two-part epoxide	Transene Epoxy 16	Non-conductive, high bond strength	Hybrid die attach	Transene Company
2. Sealants				
One-part silicone	Loctite Superflow W	Non-flow/sag	Insulation	Loctite Corporation
One-part silicone	Cho-Bond 1038	Electrically conducting	EMI/RFI Gaskets	Chomerics
One-part silicone	Dow Corning 3145RTV	No corrosive by-products	Sealing, bonding electronic components	Dow Corning

Type	Trade name	Features	Application	Manufacturer
One-part silicone	GE TRV 133	Flame retardant, non-corrosive	Electronic components	General Electric
Two-part polyurethane	Sealant U-5	Chemical and electrical stability	Cables, batteries, general electrical	Atlas Minerals and Chemicals Inc.
Two-part epoxide	Meta Seal 821	Rigid sealant	Tantalum capacitors	Mereco Products
One-part fluorosilicone	730 RTV	Resists swelling when exposed to solvents, etc.	Aerospace	Dow Corning

3. Extruding and Moulding Compounds

Type	Trade name	Features	Application	Manufacturer
ABS	Cycolac KJT	Flame retardant, high flow	TV and radio components	Borg–Warner Chemicals
Diallyl phthalate	Plaskon DAP FS-40 V.O.	Flame retarded	General electronic and electrical	Allied Chemical Corporation
Diallyl phthalate	Poly-DAP 6120	Meets various	Electrical and electronic components	US Prolam, Inc.
Epoxide	Polyset 410B	Good temperature cycling moisture resistance	Semiconductor plastic encapsulation	Dynachem Corporation
Epoxide	Polyset 450	Low stress/ high purity	Semiconductor plastic encapsulation	Dynachem Corporation
Epoxide	Polyset 110	Thermal shock resistant	Coils, bushings	Dynachem Corporation
PTFE	Hostaflow TF2053	Thermal stability	Wire insulation	American Hoechst
PTFE	Teflon 62	Thermal stability	Wire coating	DuPont
PTFE	Fluon CD125	Inert powder	Tape, high-speed cable wrap	ICI
Nylon	Amilan CM2001	Good dimensional stability and electrical properties	Electrical parts	Toray Industries
Phenolic	Polychem 129	—	Equipment cases	Buod Chemical Corporation
Phenolic	FM4008	Dimensional stability	Electro-mechanical components	Fiberite Corporation

Type	Trade name	Features	Application	Manufacturer
Polyphenylene oxide	Norylen 212	Good electrical properties	General	General Electric
Polycarbonate	Lexan 940	Flame retarded	Electrical, computers	General Electric
Polycarbonate	Panlite L-1250	High mechanical strength	Electrical parts	Teijin Chemicals Ltd.
Polyester	Hetron 92	Flame retarded	Electrical switches	Ashland Chemicals
Polyester	Pibiter NRU30	Excellent mechanical and electrical properties	Switches, connectors	Montedison
Polyethylene	Super Dylan 5003W	Long-term stability	Wire insulation	Arco/Polymer
Polyethylene	Alkathene 31/01	Good mechanical and electrical properties	Electrical parts	ICI
Polyethylene	Feretene ZF5-2400	Heat shrinkability	Miscellaneous	Montedison
Polyimide	Tribolon XT-11	High temperature stability, low wear	Connectors	Fluorocarbon
Polyimide	Polyimide 2080	Excellent thermal, mechanical and electrical properties	Fire-resistant parts	Upjohn Corporation
Polyphenylene sulphide	Ryton RID Black 5002C	Affinity for fillers, thermally stable	Electronic, electrical parts	Phillips Petroleum
Polypropylene	Plaskon FR-1080	Flame retardancy	TV components, switches	Allied Chemical Corporation
Polypropylene	Propathene GWE105	Medium flow	Thin dielectrics for cables	ICI
Polypropylene	Rexene PP41	Film	Capacitors	Rexene Polymers
Polystyrene	Stycast 350	Low electrical loss optically clear (potting)	High-frequency applications	Emerson & Cuming
Polystyrene	Esbrite 800	Thermal stability, good mechanical strength	Equipment cases	Sumitomo Chemical Company
Polysulphone	Udel P-1700	Hydrolytic stability	General electrical	Union Carbide

Type	Trade name	Features	Application	Manufacturer
Polysulphone	Udel P-6050	High impact strength	Electronic/ electrical	Union Carbide
Polyurethane	Roylar E-80	Low temperature	Cable jackets	Uniroyal
Polyvinyl chloride	Alpha 2400FR-88	Flame retarded	Electrical sleeving, harnesses	Alpha Chemicals and Plastics
Polyvinyl chloride	Blane 2150	Meets military specifications	Insulation	Blane Chemical (Reichold Chemicals Inc.)
Polyvinyl chloride	Sicron DF	BS6746(3)	Electrical insulation	Montedison
Urea formaldehyde	Plaskon Urea SMG	—	Wiring, circuit breakers	Allied Chemical Corporation

4. Casting, Encapsulating and Potting Compounds

Type	Trade name	Features	Application	Manufacturer
Epoxide	Uniset E-151	Good electrical properties	Dipcoating, potting	Amicon Corporation
Epoxide	Bacon P-75	Low density, self-extinguishing, foamed	Potting	Bacon Industries
Epoxide	Castall 248	Good adhesion to many substrates	Dip coating	Castall Inc.
Epoxide	Eccocoat D30	Impact and run resistant	Dip coating	Emerson & Cuming
Epoxide	Stycast 1263	High temperature, transparent	Casting	Emerson & Cuming
Epoxide	Stycast 1269-A	Tough, transparent	Casting	Emerson & Cuming
Epoxide	Stycast 1457FR	Flame retarded, meets MIL-I-16923	Casting	Emerson & Cuming
Epoxide	C68 Grey	High thermal conductivity low thermal expansion	Casting	Dexter/Hysol
Epoxide	Hysol C60 Std.	Excellent physical and electrical properties, semiflexible	Impregnating	Dexter/Hysol
Epoxide	Thermoset 312	Long pot life, moisture resistance	Coating, sealing	Thermoset Plastics
Epoxide	Scotchcast 241	Excellent thermal shock resistance	Encapsulation potting	3M

Type	Trade name	Features	Application	Manufacturer
Epoxide	Scotchcast 275	Fast gel, excellent chemical and moisture resistance	Coating	3M
Polyester	Stycast 40A	Good colour, low cost	Casting	Emerson & Cuming
Polyester	DuPont 49032	Good adhesion, flexibility, chemical resistance	Dip coating	DuPont
Polystyrene	Stycast 35	Low loss, low viscosity	Casting	Emerson & Cuming
Polystyrene	Stycast 350A	High dielectric constant, high thermal conductivity	Casting	Emerson & Cuming
Polyurethane	Stycast CPC-18	Excellent toughness flexibility	Casting	Emerson & Cuming
Polyurethane	CU-A3-8	Low viscosity flexibility low temperature cure	Casting	Dexter/Hysol
Silicone	3112 RTV	Low viscosity two catalyst	Encapsulation/ potting	Dow Corning
Silicone	Sylgard 184	Transparent, low viscosity	Encapsulation/ potting	Dow Corning
Silicone	Transene Hybrisil-200	One-part	Resistor encapsulation	Transene Company
Silicone	Transene Type 150	One-part, moisture resistant	Dip coating	Transene Company
Silicone	SWS T-67	Fast curing with moisture	Dip coating	SWS Silicones
Silicone	Eccosil 2CN	High temperature, clear	Encapsulation/ potting	Emerson & Cuming

INDEX